MAPPING THE NATION

Navigating Complex Challenges

Esri Press, 380 New York Street, Redlands, California 92373-8100
Copyright © 2022 Esri
All rights reserved.
22 21 20 19 18 1 2 3 4 5 6 7 8 9 10
Printed in the United States of America

ISBN: 978-1-58948-715-4

Ask for Esri Press titles at your local bookstore or order by calling 800-447-9778, or shop online at esri.com/esripress.

For purchasing and distribution options (both domestic and international), please visit esripress.esri.com.

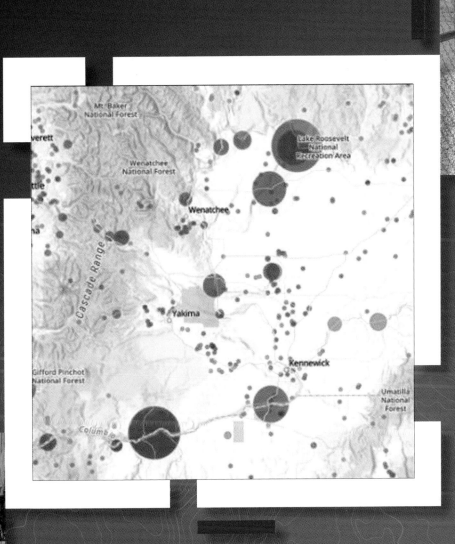

Contents

Foreword
Navigating Complex Challenges

This book contains the work of GIS users whose contributions to their organizations and broader society have increased significantly as we have faced many new challenges during the pandemic. GIS has been used to share the status of coronavirus disease 2019 (COVID-19) cases, find and serve vulnerable people, monitor spread of the virus, and efficiently deliver vaccines. During the pandemic, private- and public-sector professionals from around the nation and around the world have used GIS to address shared problems.

Mapping the Nation: Navigating Complex Challenges continues a long pattern of acknowledging our users by showcasing their work. These stories illustrate how the geographic approach and geographic information play an increasing role in better understanding and communicating complex challenges and in changing course when needed.

The title of this book is appropriate as it reflects the kinds of work our users are doing in addressing some of the greatest challenges we are facing in the world today. In a dictionary, *navigating* is defined as the methods needed to address complex challenges—to travel, especially carefully or with purpose through tough waters—using instruments or maps. The definition needs expanding, augmenting the terms "instruments or maps" and including "GIS."

Today, organizations are using the power of geographic information system (GIS) technology for all dimensions of navigation—to carefully overcome barriers that stand in the way of progress.

In the physical domain, GIS helps users navigate fleets of vehicles, coordinate mobile workforces, and rally responders safeguarding populations. It helps businesses apply spatial optimization strategies and resolve supply chain issues. GIS is also used to adroitly steer public services to achieve better outcomes. For organizational navigation, GIS allows leaders to visualize assets and operations in totality and fine-tune mobilization. It allows analysts to explore the data and simulate scenarios so planners can move forward after testing possible outcomes.

GIS makes a new kind of navigation possible to address society's many complex challenges, including growing population, climate change, social conflicts, resource shortages, and loss of biodiversity. We have all watched the severity of these problems increase as GIS users have studied and modeled the steps to a more sustainable future.

Within this book are stories about these individuals tackling the challenges of Climate, Infrastructure, Public Safety and Humanitarian Response, Defense and National Security, Conservation and Environmental Protection, Health, and Equity and Social Justice. These stories relate how geospatial technologies help organizations make and mark progress. With GIS, users have a place to start collecting and collating data, performing spatial analytics, and visualizing the multiple dimensions of a rising challenge. GIS allows us to contemplate and plan for a better future as it guides incremental daily improvements.

GIS users apply the geographic approach—a way of thinking and problem solving that integrates geographic information to understand and manage a variety of challenges. The geographic approach also allows us to apply this knowledge to design and planning—to remap our world.

My colleagues and I are constantly amazed by the depth of GIS activity and how its use brings measurable improvements to organizations and communities. The work and ideas of GIS professionals are helping advance science in so many ways: to design with nature in mind, to make communities more livable and efficient, to improve public safety, to secure national defense, to protect natural spaces, to ensure human health, and to mitigate social conflicts.

Our users are navigating tough challenges—by traveling, especially carefully and with purpose—using GIS to overcome any barriers.

Warm regards,

Jack Dangermond

Introduction
GIS Components to Tackle Complex Challenges

Around the world, organizations are applying apps, drones, artificial intelligence (AI), machine learning, and the Internet of Things (IoT) to populate a real-time view of assets and people on a shared map to produce digital transformations and address the challenges our nation faces.

A modern GIS combines data collection, analysis, and sharing to achieve operational intelligence. With these tools, organizations gain an edge on complicated challenges because they can see trouble coming and manage decisively around it. For cities, regions, and nations, these same tools help deliver services more efficiently and cost-effectively, increasing residential awareness through maps to ensure that all communities are well served.

Decision-makers and operations officials require a real-time management tool to see current conditions so they can make better decisions about resources, people, and assets.

And these leaders need to see things not just as they are now, but how the situation is likely to shift in an hour, a day, or a week. They need a tool that provides a visual means to understand the full scenario, and the way its context affects—and is affected by—the movement of people and assets. They need a technological solution to see what's happening and query across space and time to make crucial resource management decisions, acting proactively rather than reactively.

Using GIS technology can help you accomplish these interwoven objectives, with a suite of tightly integrated tools that can help you achieve several goals simultaneously:

Collect, Analyze, Share—use purpose-built, location-based apps to collect data and optimize the efficiency of field activities. Then use GIS in the office to store, map, and analyze data points to see challenges in their totality. The data, map products, and analytical results can all be shared internally, across departments, with partners, and back and forth to the field.

Spatially Enable Operations—operational awareness delivered by GIS lets managers see what is happening, track mobile employees, reduce mileage and fuel costs, save time and wear and tear on vehicles, and push routes and directions directly to the field to simplify communications and speed up services.

Achieve Real-Time Intelligence—ingest massive volumes of real-time data feeds and perform fast queries and analysis to understand movement and change. Agencies are increasingly constructing digital twins, which combine a 3D model with IoT sensor data to see activity and the workings of the urban environment in real time. This real-time awareness allows users to see such things as bottlenecks in traffic as they occur, where to respond in an emergency or crisis, and where service gaps exist.

Integrate Important Business Systems—GIS provides a common meeting ground for other enterprise systems through the powerful factor of location. Through this primary attribute, contained in an estimated 80 percent of all records, relationships are established, and data from multiple systems can be seen, queried, and acted on.

See the Situation with Clarity—because GIS contains tools to understand people, things, processes, and situations, it provides unique context. It's about being empowered through your devices, giving everyone fast and correct answers, and getting to consensus quickly.

Bring Stakeholders Together for Shared Solutions— GIS workflows underpin good decision-making by helping users analyze the data at hand, target the workforce to take action, and then monitor progress.

GIS can bring data together from disparate sources, with smart maps offering a comprehensive view of what's happening now. Weather reports, incidents, construction, mobility—it all comes together to provide full awareness of current conditions and of ways to improve them. With access to historical data, trends and patterns can be discovered to guide decisions. Armed with location intelligence, management can proceed at peak capacity, meet high performance levels, and achieve service excellence with the goal of creating a sustainable environment.

GIS manages this level of complexity while bringing visibility to problems and the awareness of what it takes to progress toward meeting the sustainability goals that define a resilient city—and nation—for the 21st century, from providing for the essentials to supporting a comfortable and vibrant way of life.

An Ecosystem of Open Data

While Esri® software started as mapping and spatial analysis tools for consulting projects, efforts soon shifted to developing a general-purpose GIS product to sustain a wider community. This shift changed the focus of Esri from being a consulting service-based organization to a software company.

Over time, through design, GIS evolved to become a standard commercial off-the-shelf (COTS) product that implements a variety of open strategies. This change provided Esri customers with better reliability, security, and independence. It also allowed Esri to scale its technical support and provide better quality, consistency, and strong documentation.

Today, Esri's technology, ArcGIS®, is a powerful open and interoperable platform that is used by hundreds of thousands of organizations. It's designed to support individual projects and to scale to entire enterprise systems.

Among the benefits of ArcGIS is an open, supported, and well-engineered architecture that includes rich functionality for data capture, data management, mapping, 3D visualization, and spatial analysis.

Thousands of software developers have embraced the open ArcGIS system to build and deliver enterprise applications and custom systems, illustrating the success of this flexible approach.

For much of its history, GIS has been constrained by both a lack of data and the time-consuming tasks required to gather and manage it. Today, with open data repositories such as ArcGIS® Living Atlas of the World, GIS users simply connect to a trove of data and can unleash their creativity to create maps and apps built on a backbone of readily accessible basemaps and data layers.

GIS users have a wide range of application frameworks to deliver apps tailored to user workflows. Apps have seen wide adoption, which leads to the capture of more and better data in a perpetual cycle of improving inputs. Apps also work well as a coordination mechanism, to get a dispersed team, especially in the field, to work together as one.

And now using Web GIS, organizations can share data as services to connect apps across organizations, with partners, or with the public. Interactive maps that show conditions, and even the live locations of participants, foster collaborative action.

Another data advancement is the use of cloud computing coupled with artificial intelligence (AI) to process huge volumes of remotely sensed data. With data science, these large volumes are used to detect patterns and make predictions. Many forms of spatial analysis help determine the areas best suited for a specific service or the priority places to address vulnerabilities, but with big data it's about returning more nuanced answers.

Esri continues to work with users to push new GIS functionality. Some of Esri's current work is focused on advancing analytics and visualization using big data, real-time Internet of Things (IoT) data, and new distributed GIS architectures to break down barriers for sharing geographic knowledge and encouraging citizen engagement.

Esri is passionate about the ability of GIS to address complex problems and is constantly striving to increase the power of GIS and make it easier to use.

Esri's open platform approach provides flexible, configurable, customizable, scalable, interoperable, and easily deployable technology. Esri's ongoing goal is to enable its users and partners to apply their creativity and skills to inform decision-making while ensuring that GIS innovation and application continue to move forward along their many pathways.

Esri Provides Open Access to Key Federal Geospatial Data

Free Open Data, Data Downloads, and Web Services

Working with federal agencies, Esri is opening access in ArcGIS® Online to dozens of high-priority, high-demand national data layers, often referred to as A-16 data after the US White House Office of Management and Budget's Circular A-16 that detailed the themes of the national geospatial data asset portfolio. These layers include cartographic boundary files and demographic data from the US Census Bureau; National Agriculture Imagery Program (NAIP) data from the US Department of Agriculture (USDA); the National Inventory of Dams (NID) database, managed by the US Army Corps of

The National Land Cover Database was used to produce this 3D map of Anchorage, Alaska.

Engineers; and the National Land Cover Database (NLCD), which is the result of federal agencies working together to create up-to-date, consistent land-cover products for the entire country.

Esri is making three significant enhancements that will broaden access to A-16 data. First, this data collection is now publicly accessible under an open commons license agreement, making it free for anyone to use. Second, the data is available for download in a variety of interoperable formats. And third, it is served as both Esri and Open Geospatial Consortium Inc. (OGC)-compliant services, meaning it is ready to incorporate into apps.

These open data services are provided through ArcGIS Living Atlas of the World, which significantly expands their visibility. Additionally, the new open-access license, along with added technical options for using the data services, extends their reach beyond the Esri community. This supports the vision of the Geospatial Data Act of 2018 (GDA), which seeks to empower more people to use geospatial data and breaks down more barriers to interoperability.

The Data Is There—Now It's Time to Open It Up

Globally, the trend is toward making data more open, accessible, and reusable. The number of public open data sites provided by national agencies, as well as state and local governments around the world, has skyrocketed, with more than 11,000 powered by ArcGIS alone.

Recent legislation in the United States, including the GDA and the Open, Public, Electronic, and Necessary (OPEN) Government Data Act of 2018, is pushing this movement forward at the national level. It calls on all agencies that serve National Geospatial Data Assets (NGDA)—the priority national data layers—to expand standards-based open access. The overriding intent is to foster greater public- and private-sector use of federal data to encourage innovation and data-driven decision-making.

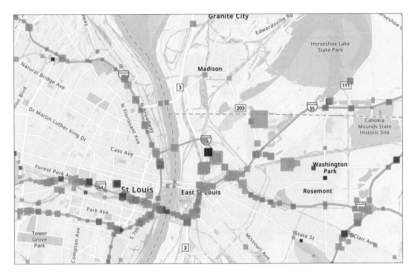

A-16 geospatial data layers cover features such as bridge conditions, shown here for Saint Louis, Missouri.

US federal agencies have excelled at creating and managing critical geospatial data to meet their own needs. Disseminating the data widely and making it available as accessible, usable services are more recent endeavors. Successful examples of open data providers at the national level exist, however. For example, the Department of Homeland Security's Homeland Infrastructure Foundation-Level Data (HIFLD) program works with multiple federal agencies to curate and distribute geospatial data services. Demand for these web services is growing, with usage up 75 percent from 2019 to 2020. So the desire for open federal data is there.

But this is just the tip of the iceberg. Vast amounts of federal data exist, although many data owners have yet to modernize their data management and dissemination infrastructure. Increased collaboration among Esri and federal agencies will accelerate the pace for serving open data.

Expanding access to federal data by making it public in ArcGIS Living Atlas empowers a wider audience to address challenges such as climate change, disaster relief, broadband access, and racial equity. Further, offering that data as services makes it readily usable by not only GIS professionals but also by web-savvy developers, scientific audiences, and even citizen scientists and app developers.

This is the only way to satisfy the data demands of today and tomorrow and increase the value of the investments made in geospatial data.

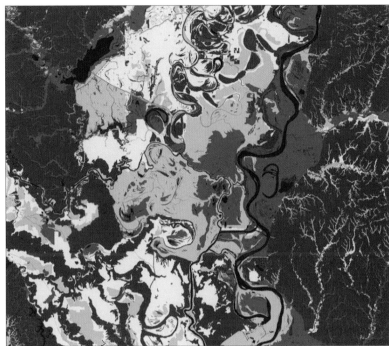

The Soil Survey Geographic Database (SSURGO) is available through ArcGIS Living Atlas and can be used to see the effects of long-term flooding.

New Levels of Inclusiveness and Interoperability

To help address current data access and use challenges, Esri is working collaboratively with federal agencies to amplify their work by increasing the visibility of their data and delivering it in app-ready formats.

A new collection of federal maps and apps is now available in a curated, Esri-managed group in ArcGIS Online called the US Federal Maps and Apps organization. To government users, this collection of data is known as A-16. To everyone else, this is a broad collection of nearly 100 national geospatial datasets on dams, bridges, federal public lands, runways, rail nodes, and other elements. The collection serves the Esri and geospatial user communities as well as the open-source and web developer communities.

To enable enhanced open access to this collection, Esri is changing both the terms of use and technical parameters of the data. The collection is now available under the Creative Commons Attribution 4.0 International (CC BY 4.0) license, meaning anyone can use, share, and adapt the data with proper attribution. This considerable policy update removes a historic barrier to access.

On the technical side, Esri is making cached and direct connections to public services within the federal government. The layers can be exported to multiple open formats, such as a shapefile, CSV, KML, and GeoJSON, to perform further analysis. They are also accessible as OGC-compliant standards, including Web Map Tile Service (WMTS) and Web Feature Service (WFS) files, as well as Esri formats to support interactive exploration, visualization, and analysis. The geospatial layers are updated regularly and reference the federal agencies' metadata.

These advancements in open access to federal data demonstrate a new level of inclusiveness and interoperability across the geospatial community. Nadine Alameh, the OGC's chief executive officer, welcomed this development.

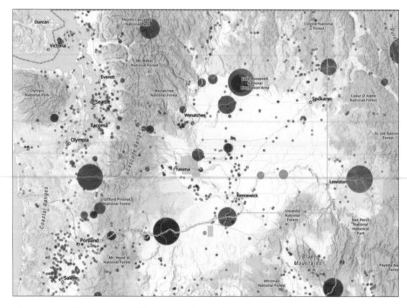

The National Inventory of Dams database is a high-priority national dataset, used here to map dams in parts of Washington and Oregon.

"I'm happy to see a practical, operational example of how open standards and services can advance the GDA objectives," she said. "Thank you, Esri, for making this government data available not only to its vast geospatial community but also anyone through open OGC services, enabling further interoperability and innovation."

Benefits That Cut across Governments and Sectors

This collaborative effort implements a "create once, use many times" philosophy.

Because the federal government produces and manages key data, the pressure is reduced on other organizations to engage in these time-consuming and cost-intensive tasks. Once that data is created and shared, it is advantageous for users to have many pathways to find it. Although federal agencies can publish their data on their own websites, put it on thematic or regional portals, and post it to the GeoPlatform run by the Federal Geographic Data Committee (FGDC), making the data available through Esri technology amplifies its exposure and increases the likelihood that it will be discovered and used.

According to Ivan B. DeLoatch, executive director of the FGDC, developing public-private partnerships is critical to advancing the United States' National Spatial Data Infrastructure.

"The exploration of a public partnership [between] Esri and the FGDC community will be an exciting opportunity to support the ongoing development of our National Geospatial Data Assets, or A-16 datasets," DeLoatch said. "This will encourage the use of open data and standards, as well as data-sharing initiatives with stakeholders, to address national challenges and priorities."

For example, the data needs of state and local governments do not stop at their borders. They often rely on federal data to add broader perspectives to local projects and initiatives. The National States Geographic Information Council (NSGIC) has long advocated for the federal government and organizations across the country to provide openly accessible and interoperable geospatial services. And Frank Winters, current president of the NSGIC and executive director of the New York

State Geospatial Advisory Council, commended this effort for advancing that vision.

"We need to maximize the reach of geospatial data, and that means we should think of everyone as a decision-maker," said Winters. "When our data reaches thousands or millions of people—to help them decide where to live, buy property, go to school, or start a business—our impact can scale to a point where it moves the dial of our economy. Removing barriers to high-quality spatial data, time and again, results in people using data in ways I could never predict, making interesting and meaningful contributions to their communities."

In the commercial sector, for instance, large companies with assets and interests spread throughout the United States need data that cuts across cities, states, and regions. Having widespread high-quality national datasets that are easy to access and available as services benefits countless businesses, from national retailers and insurance companies to railroads and utilities. It saves them the time and money required

to gather and process the data themselves and enables them to carry out important operations more quickly and cost effectively, such as conducting risk assessments, doing environmental screenings to evaluate the feasibility of getting a project permitted, and analyzing growth opportunities.

Continuing to Elevate Federal Data

Geospatial data is critical to understanding and meeting the challenges that continually crop up all over the world. Esri is committed to collaborating with federal government organizations and agencies to make their spatial data easy to access and use so that it can better serve the geospatial community, developers, scientists, and, ultimately, the public.

It is Esri's goal to build on the great content that federal agencies already collect and bring it to life through well-documented, open, and interoperable services and downloads. This cooperative effort will elevate the value of that data through increased usage, fuel the development of problem-solving apps across all sectors, and spur new innovations.

Expanding access to federal data empowers more people to address challenges such as racial equity—for example, by building dot density maps that highlight different racial and ethnic groups.

To Better Understand the Earth

By Monica Pratt, head of Esri Publications

With the release of ecological coastal units (ECUs), another tool for understanding the earth is available. Created for mapping and analyzing the world's coastal regions, ECUs are the most recent development in an ongoing effort to create a global framework for understanding ecosystems on land, ocean, and shoreline that provide the services that support life on Earth. GIS plays a vital role in building, visualizing, and using this framework for better understanding ecosystems and preserving them.

Completely mapping the earth in a standardized and rigorous manner is essential for tackling global challenges, such as climate change and the preservation of biodiversity, which are becoming more pressing concerns. Accurate geospatial information will make it possible for scientists, land managers, conservationists, developers, and the public to understand the current state of ecosystems and contribute to a more profound understanding of how they are changing. This data is vital to identify areas that must be protected and others that need to be managed in a more sustainable fashion.

Although an abundance of ecosystem data has been gathered, it has not been available at a global extent created with standardized scale and quality. Previously, no organization had undertaken ecological base mapping on a global scale using standardized units and terminology. Without a shared language for reference, it can be difficult or impossible to develop understanding.

The need for the first comprehensive maps that define all of Earth's ecosystems was recognized and supported by the Group on Earth Observations (GEO), a partnership of more than 100 national governments and more than 100 participating organizations that work to create innovative solutions to global challenges that transcend national boundaries.

GEO's Biodiversity Observation Network (GEO BON) and GEO Global Ecosystem Initiative (GEO ECO), GEO's initiative to map and monitor global ecosystems, are working on a grand scale to map and understand global challenges. A high-resolution, standardized, and practical map of global ecosystems for terrestrial, marine, and freshwater environments is an initiative of GEO ECO. Roger Sayre, senior scientist for ecosystems at the US Geological Survey (USGS), was asked to lead the US portion of this initiative and a team composed of scientists from USGS and various academic institutions, as well as geographers, analysts, cartographers, and software developers from Esri.

Beginning on Land

In 2013, USGS and Esri began this formidable undertaking by mapping ecological land units (ELUs). They built a GIS of world landforms using a 250 by 250 m grid that contained 3.5 billion cells. A collection of historical maps was used to validate and corroborate this data. The USGS and Esri vetted each of the input datasets, ensuring that each was the best available at the time, and evaluating each for data quality.

More than 105,000 ecological facets describing those cells were identified. These facets represent unique combinations of land cover, lithology, landform type, and climate. The facets were integrated based on different aspects of a terrestrial ecosystem—climate regime, soil moisture, geology, landform, and the organisms present—and aggregated into 4,000 ELUs, using GIS to create a geographic set of ecosystem areas that can be visualized and analyzed. ELUs were first made publicly available in 2015.

ELUs were the beginning of a global framework. They fill the gap between micro and macro views with medium-scale data that is useful for studying ecological diversity, rarity, and evolutionary isolation. Data at this scale is especially valuable for conservation and preservation planning. By establishing common units and shared terminology, ELUs enable the creation of greater collaboration.

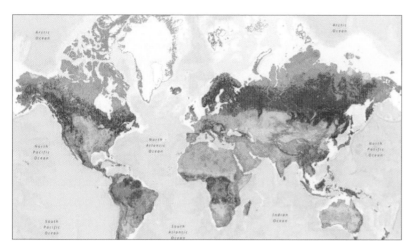

The creation of ELUs was the beginning of a global framework for studying ecological diversity, rarity, and evolutionary isolation.

Mapping of the ELUs was based on structural elements, rather than biological assemblages. Once species distribution data is incorporated, researchers will be able to explore questions about species range and distribution across the world.

A Framework for Understanding Oceans

The work of the initiative moved from Earth's landmass to its oceans with the development of ecological marine units (EMUs). As the repository of almost all water on Earth, oceans

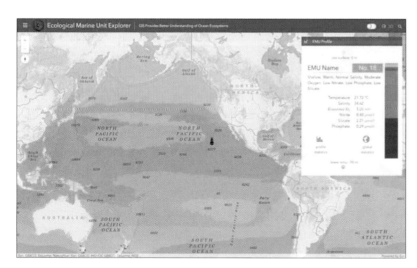

EMUs delineate physically and chemically distinct regions of the open ocean and note variations in temperature, salinity, oxygen, and nutrients.

are critical to the well-being of the planet and its inhabitants because of their impact on climate and weather and their role as a major source of food. Yet despite this undeniable importance, it is estimated that only 10 percent of oceans have been explored.

To ameliorate this situation, a public-private partnership built the EMUs. The partnership was led by Esri and USGS in collaboration with NatureServe, the Marine Conservation Institute, the University of Auckland (New Zealand), GRID-Arendal (Norway), Duke University, the Woods Hole Oceanographic Institution, the National Institute of Water and Atmospheric Research (NIWA), the US National Oceanic and Atmospheric Administration (NOAA), and the US National Aeronautics and Space Administration (NASA).

The project establishes a 3D point mesh framework spanning 52 million points and amassed global measurements of six key variables of the ocean's water column over a 50-year period. A big data project, the creation of EMUs represents data aggregation and computation on an unprecedented scale. Climatology data was extracted at one-quarter degree by one-quarter degree intervals (or approximately 27 by 27 km as measured at the equator) at variable depths before being spatially analyzed and clustered using a multivariate statistical method, and then verified by leading oceanographers.

Released in 2016, EMUs delineate physically and chemically distinct regions of the open ocean and note variations in temperature, salinity, oxygen, and nutrients. They are a standardized, rigorous, and ecologically meaningful set of ocean ecosystem units that may be used as a basemap for climate change impact studies, biodiversity priority setting, economic and social valuation studies, research, and marine spatial planning.

In describing the unique value of EMUs, Esri chief scientist Dawn Wright, who led Esri's efforts on the project, said, "The strength of EMUs is that they differ from existing maps of marine ecoregions or biogeographic realms by being globally comprehensive, quantitatively data driven, and truly 3D."

This is data that can help inform the actions of conservation-minded organizations, academic institutions, and scientists who are working to preserve marine environments. NOAA's World Ocean Atlas is the primary data source for EMUs. They are easily accessed from Ecological Marine Unit Explorer apps via the web browser and mobile devices and can be downloaded by joining the ArcGIS Online Ecological Marine Units group.

Between Land and Sea

Although the 27 by 27 km resolution of EMUs is fine for open ocean data, the much higher 1 km resolution was needed for coastal areas. The team that produced the EMUs developed a separate and independent effort to delineate global ECUs in 2018.

The ECU for Fiji, accessed via the Global Islands Explorer.

This new map covers the unique and extensive coastal region, which differs greatly from terrestrial and open ocean environments. In addition to the coastline itself, the global coastal zone includes the terrestrial ecosystems and marine ecosystems on either side of the coastline.

Understanding coastal zone ecosystems is critical to sustainability efforts and can help mitigate the environmental, economic, and social effects of climate change. Understanding coastal ecosystems is essential for meeting the United Nations (UN) 2030 Agenda for Sustainable Development and its 17 Sustainable Development Goals (SDGs), Specifically, ECUs support the UN's SDG 14: Life Below Water goal, which calls for the sustainable management and protection of marine and coastal ecosystems and the conservation of at least 10 percent of coastal and marine areas. Achieving this goal requires a comprehensive inventory of coastal ecosystems and a strategy for determining the 10 percent of ecosystems to be preserved.

According to estimates by the UN, nearly 2.4 billion people, or about 40 percent of the world's population, currently live within 100 km of the coast, where they are at risk from flooding caused by sea level rise. Understanding coastal regions will be necessary to limit the social, economic, and environmental costs of sea level rise that analysis has shown will likely increase within the next decade.

Creating ECUs

To build ECUs, the team took a coastline segmentation approach to analyze the differences in coastline properties. The initial step in the process was the development of a brand-new 30 m resolution image-derived map of Earth's coastlines. This effort was necessary because the coastline characterization then available was inadequate in terms of accuracy, spatial resolution, comprehensiveness, and the visual fit of the coastline vectors to the shore of existing noncommercial global shoreline datasets.

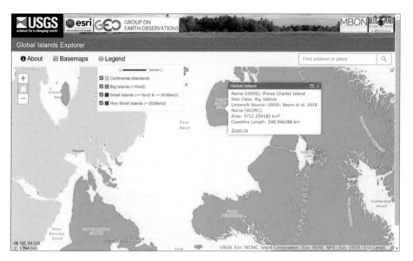

ECUs cover the unique and extensive coastal region, including Prince Charles Island *(upper right)*. ECU data is available from the USGS Global Islands Explorer.

These segments served as the basis for subsequent analysis. Each segment was attributed with 10 variables describing the ecological settings for that segment. Those segments were classified into 81,000 unique coastal segment units (CSUs) based on the Coastal and Marine Ecological Classification Standard (CMECS), a framework for organizing information about coasts and oceans and their associated living systems which is endorsed by the FGDC.

CSUs were statistically clustered into 16 groups that are globally similar as described by the 10 variables. This process yielded ECUs, standardized data that facilitates comparison of coastal ecosystems on an apples-to-apples basis and encourages collaboration between organizations.

A Valuable Resource

Although created as inputs for generating ECUs, CSUs are valuable in efforts to understand coastal ecosystems. CSUs provide a descriptive global inventory of coastal ecological settings that have management utility at local scales because they use data at the highest available spatial resolution. ECUs are exploratory and preliminary units that have utility for global assessments, especially for the UN's 30×30 goal to preserve 30 percent of land and 30 percent of the oceans by 2030 as well as the UN SDGs.

ECUs join ELUs and EMUs in supporting the creation of a set of standardized, rigorous, and ecologically meaningful maps. All have been produced using an objective and repeatable approach that uses big data techniques to synthesize and classify ecologically important data layers into distinctive and meaningful georeferenced units.

The next steps in creating this framework for understanding Earth will be the creation of global ecological freshwater units (EFUs) and, possibly, ecological benthic units (EBUs).

These global ecological units are powerful and valuable to a much broader universe. They help promote a more profound understanding of the earth that has potential for informing more sustainable decisions by governments, the private sector, and nongovernmental organizations.

In a 2019 interview, Sayre said, "We needed a coastline vector as the spatial backbone for this project, so we went into the cloud and extracted a brand-new global shoreline from 2014 Landsat imagery. Looking at this new global shoreline, it occurred to us that since every landmass on the planet is an island, no matter how big it is, we were sitting on a new global islands database." The 2014 annual composite Landsat imagery base was used because it dilutes the effect of clouds in imagery.

The result was the global shoreline vector (GSV) and the global island dataset derived from GSV through the application of polygon topology. GSV was developed using a semiautomated supervised classification approach using big data techniques. Because Landsat imagery was already available in the cloud, the need to house and manage vast amounts of imagery was eliminated.

This process captured data for hundreds of thousands of islands that were missing from previously created datasets. By developing the Global Islands Explorer (GIE), USGS and Esri made this new global data on islands accessible. Using this tool, 340,691 islands can be visualized and queried.

From GSV, four million, 1 km segments were generated.

UN Sustainable Development Goals Put a Global Spotlight on Local Action

Climate change, inequality, and poverty are deep, interconnected challenges our world faces. To solve issues impacting both people and planet, the UN General Assembly designed the Sustainable Development Goals. The SDGs comprise 17 objectives aimed at creating a safer, healthier, and more sustainable future worldwide by the year 2030.

To date, 193 countries have formally adopted the SDGs, demonstrating a nearly total dedication of world leaders to take the actions necessary to achieve these common goals. Many are using a GIS to share SDG data, stimulate policy formulation, and shape decision-making.

Although the SDGs are giving national leaders a set of focused goals and GIS is providing the technology foundation, it will take hard work at local levels to achieve progress. Here we profile devoted people and organizations taking action and explore their approaches.

SUSTAINABLE DEVELOPMENT GOALS

The 17 SDGs represent the common goals of 193 countries to make a more sustainable future.

Feeding America Riverside | San Bernardino

The community-based organization Feeding America Riverside | San Bernardino (FARSB), a member of the Feeding America national network of food banks, fights hunger in the Southern California counties of Riverside and San Bernardino. More than 800,000 people live below the poverty line in the group's 27,000 sq. mi. service area. Half of those below the poverty line experience food insecurity, and 51 percent are children. FARSB works to alleviate hunger, which ties in directly with SDG 2: Zero Hunger.

GIS helps food banks plan food distribution, but FARSB wants to take mapping and spatial analysis further—addressing hunger at the source. This includes closely examining causes of food insecurity such as poverty, unemployment, and poor educational opportunities. FARSB recognizes poor nutrition as an early precursor to other problems, often leading to chronic illness and increased health-care costs, which in turn lead to difficulty in securing employment, further reducing income and increasing food insecurity. Targeted approaches aim to break the cycle.

"GIS technology allows FARSB to explore overlapping layers of need and identify the 'where.' Where can we deploy limited resources for maximum effect? Where are the harmful patterns we can interrupt in our community?" said Vanesa Mercado, director of programs, Feeding America Riverside.

Since the onset of COVID-19, FARSB has used the GIS-based Homebound Emergency Relief Outreach (HERO) program to reach seniors and people with disabilities who are further isolated by the global pandemic. This effort proves particularly critical at a time when the pandemic has increased the local food insecurity rate by an estimated 48 percent.

FARSB uses GIS to map the locations of homebound individuals and pair them with nearby volunteer food delivery drivers. Additionally, HERO program staff use GIS to identify which distribution centers are closest to volunteers and clients. Individual maps are aggregated to provide stakeholders with an understanding of the full scope of the project and expose concentrations of food insecurity to guide future outreach.

Direct Relief

Founded in 1948, Direct Relief is a global humanitarian organization dedicated to improving the health outcomes of disadvantaged people and those affected by emergencies. To ensure availability of essential medical resources, the group relies on a network of more than 4,000 health-care providers in more than 120 countries.

In support of SDG 3: Good Health and Well-Being, Direct Relief equips health professionals with lifesaving medical supplies to care for the world's most vulnerable people. A particular focus, maternal and child health, seeks to protect women and babies through critical periods of pregnancy and childbirth, including birth-related injuries.

One birth-related injury, obstetric fistula—a hole in the birth canal caused by prolonged and obstructed labor—often results in a stillborn baby and leaves women isolated and stigmatized. This preventable condition was once experienced worldwide, but due to improvements in post birth health care, incident rates are lower in most places except for disadvantaged areas of Africa and Asia.

To support the United Nations' aim of eliminating obstetric fistula by 2030, Direct Relief partnered with the Fistula Foundation to create the Global Fistula Hub. Built with ArcGIS® Hub from Esri, the site provides open data, data analytics, data visualizations, and information about industry events, as well as a venue for connection and collaboration.

"We want to promote evidence-based maternal child health-care practices and feature the work being done by the global community to remove the threat of obstetric fistula for every woman, everywhere," said Jessica White, GIS project specialist, Direct Relief.

UN Sustainable Development Solutions Network

To drive the use of modern technological and scientific advances for practical solutions, the Sustainable Development Solutions Network (SDSN) was created under the auspices of the UN in 2012. The SDSN coordinates with UN agencies, the private sector, and related organizations on various initiatives, including SDG implementation. In this role, the SDSN complements the UN's work by calling attention to related issues.

The SDSN partnered with Esri and the National Geographic Society to create SDGs Today, a global hub site for real-time SDGs and data. Policy makers, researchers, students, and the public can access the hub site for accurate, real-time spatial data on the status of the SDGs in all participating areas. The hub site includes a portal for each of the SDGs, where users can access datasets and descriptions; data visualizations; maps; and supporting resources, including online courses and lessons, to integrate the data into their own projects.

One example is a dataset developed by WorldPop, an organization that specializes in mapping demographics in low- and middle-income economies. The dataset addresses SDG 3: Good Health and Well-Being by focusing on the percentage of sub-Saharan women of childbearing age who live within two hours' travel time of a health-care facility. WorldPop provides the grid-based population data needed to calculate accurate estimates that support expanded access to health care.

SDSN, Esri, and WorldPop are part of the POPGRID Data Collaborative, a consortium that encourages innovation and cooperation to deliver accurate geo-referenced population, settlement, and infrastructure information. "To curate timely and spatially disaggregated data to move projects forward, and to help policy makers make more informed decisions for today and for the future, is not an easy task," said Maryam Rabiee, manager TReNDS and SDGs Today at the United Nations. "We can't do it alone."

SDSN also publishes sustainable development reports annually to assess each participating country's SDG progress. To make these reports more accessible, staff created an ArcGIS® StoryMaps℠ story and an interactive dashboard. Site visitors can explore the map interface to learn about a country or SDG.

To make GIS data easier to use and understand in global SDG work, SDSN has been developing an online GIS course to include training sessions for SDSN members and education modules for universities so students can enhance their GIS skills. Toward that goal, SDSN is also working with Esri on the ArcGIS StoryMaps Competition for the Sustainable Development Goals. The contest encourages students and professionals to tell compelling stories using GIS tools and SDG data.

W.K. Kellogg Foundation

The W.K. Kellogg Foundation (WKKF) was founded in 1930 as an independent, private foundation by breakfast cereal innovator and entrepreneur Will Keith Kellogg. WKKF works with communities to help vulnerable children realize their full potential in school, work, and life. Although the Kellogg Foundation supports efforts across many of the SDGs, SDG 10: Reduced Inequalities is a particular focus of the foundation.

In recent years, the United Nations Global Geospatial Information Management Secretariat (UN-GGIM), Esri, and other organizations have worked together to establish the International Geospatial Information Framework, (IGIF). The IGIF makes it easier for countries or regions to establish a shared geospatial infrastructure to support SDG change efforts and track progress toward the goals. This framework has been adopted and is in use in several countries, but many developing countries lack the tools and resources to support collaborative change efforts, and to track progress toward the goals.

WKKF wants this interconnected framework to facilitate the flow of knowledge and learning. In that spirit, WKKF worked with local partners to implement a geospatial data hub to support communities in the Chiapas and Campeche regions of Mexico. As founder of WKKF, Kellogg believed that cooperative planning, intelligent study, and group action are essential steps to create long-lasting change in communities. This fundamental "theory of change" helps guide the work of the organization today. Through collaborative partnerships and a supporting grant, WKKF is working with Haiti, the UN, Esri, and local partners to bring this powerful, enabling infrastructure to support the SDG and development priorities in Haiti.

"The boots on the ground efforts of local organizations require resources, planning, and geospatial tools, but more than anything, it requires the commitment and perseverance of steadfast individuals who are determined to create change at the community level," said Ross Comstock, vice president of Information Systems and Technology at WKKF. "We are listening to changemakers—the people and organizations working together on positive change. If we put the needs of changemakers at the center of our efforts, could we help to accelerate the pace of positive change?"

The Kellogg Foundation promotes this effort of listening to, learning from, and supporting changemakers as one way that philanthropy can support positive change, and progress toward those goals.

While working on the 17 global SDGs, staff at local organizations need resources, including geospatial tools for planning, monitoring progress, and sharing information. What's needed more than anything else to reach this safer, healthier, and more sustainable future are individuals and leaders determined to create change at the community level.

Vanesa Mercado
Director of Programs,
Feeding America Riverside

Jessica White
GIS project specialist,
Direct Relief

Maryam Rabiee
Manager of the SDSN's Thematic Research Network on Data and Statistics (TReNDS) and the SDGs Today program

Ross Comstock
Vice President of Information Systems and Technology at W.K. Kellogg Foundation

The York Region Administrative Centre, designed by Canadian architect Douglas Cardinal, was completed in 1993.

York Region Pioneers Next-Level Local Government Collaboration

The Regional Municipality of York, or York Region—Canada's fastest growing municipality—has pioneered a collaborative, data-driven approach to local government that is gaining global esteem. The digital transformation provides the means for local governments across the rapidly changing region to share data and apps and work together to better understand and serve residents.

Elements of York Region's YorkInfo Partnership include direct data connections, app sharing, tools and computer code collaboration, and an academy with workshops for analytics professionals to improve their skills.

The idea to create a data cooperative to achieve efficiencies and eliminate data duplication isn't new.

"We found meeting minutes in our archives from 1977 that said we need to find a way to set up a utility for sharing our data," said Jeff Lamb, manager of partnerships at York Region. "The concept has been there, and the technology caught up."

Other local governments marvel at how resolving data-sharing bottlenecks has allowed York Region to apply advanced analytics to improve decision-making.

"We started thinking about what keeps senior management up at night," said John Houweling, director of data analytics and visualization at York Region. "It's about business drivers and problem-solving."

Moving from Vision to a Way of Working

York Region has grown steadily to become the sixth-largest municipality in Canada, with almost 1.2 million residents. With the constant addition of jobs, people, homes, and infrastructure, the local government relies on close collaboration to keep data current.

York Region combines a strong regional government with nine municipalities that collectively oversee transportation; mass transit; water, wastewater, and solid waste; policing; emergency services; housing; human services; and planning. Most of these services are managed in part by a GIS, which applies a location context to guide operations.

"Everything has either an address or it's happening on a property, so it can be attached to our address points, parcels, and roads," said Laura Thomas, GIS analyst at York Region. "We take a lot of business data and bring it into the spatial world."

Although an enterprise technology architecture connects the systems that manage each service, the region identified a need for people to become better connected as well.

"We need everyone to work together, because something like 60 percent of the data we rely on for regional programs comes from local municipalities," Lamb said. "It takes dedicated work to get people at the table and keep them

there. Over time, we've found ways to break down barriers at all levels."

Now, the data sharing has become routine.

"All the collaboration and partnerships are part of our ordinary daily work," Thomas said. "It's only when I go to a conference that I realize it's a bit different than what everyone else is doing."

Combining and Connecting Data

The partnership formed in 1996, with York Region in the middle as the hub and municipalities as the spokes. Originally, it was purely about the exchange of GIS knowledge and data, and the exchanges were mostly between the region and each municipality individually. In 2013, the first umbrella data-sharing agreement eliminated the back-and-forth of every data request, meaning that when a request was received, the data was simply delivered.

"The partnership agreement proved pivotal to making everyone feel more connected," Lamb said. "Before that agreement, technology wouldn't have solved the human barriers."

One of the first regional data efforts was to produce the All Pipes database, which aggregated infrastructure data from each of the nine municipalities. York Region provides water and wastewater distribution pipes, and each municipality is responsible for transmission pipes to homes or businesses.

"We created data standards and a data model to bring all the data together," Houweling said. "We're now using that approach to create common data for different things—all streets, all parks, all whatever."

There's also an ongoing collaboration to capture the construction plans of each municipality using a common viewer. It will allow everyone to see what their neighbors have planned or are thinking of doing. This visibility starts conversations to prevent disruptions or to combine efforts for cost reductions.

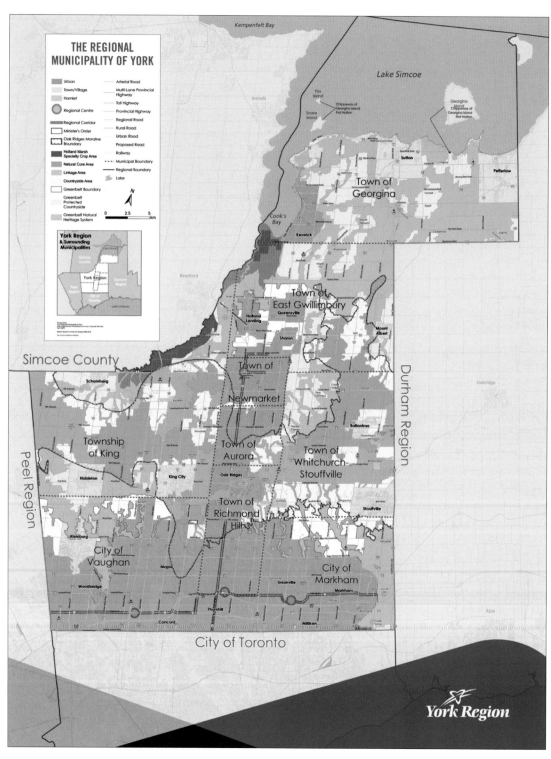

York Region combines a strong regional government with nine municipalities: the City of Markham, City of Vaughan, Town of Richmond Hill, Town of Newmarket, Town of Aurora, Town of Georgina, Town of Whitchurch-Stouffville, Town of East Gwillimbury, and Township of King.

York Region applies a distributed collaboration approach, using a combination of ArcGIS technology to seamlessly share GIS content, including maps, apps, and layers, across partner organizations.

"The beauty of the technology is that we can connect and put our live production data up there and forget about it," Thomas said. "Our processes are running, and the data stays connected and alive."

The burgeoning Town of East Gwillimbury, one of the region's nine municipalities, provided a proving ground for this next level of real-time data sharing. The town has traditionally issued 150 building permits per year. In 2020, that total jumped to more than 1,000.

"It's almost ridiculous how fast the town is growing," Thomas said. "We need to know the occupancy of a new subdivision—and when people are moving in—as quickly as possible to provide new residents with services."

Data comes in quickly to the co-op through a combination of GIS apps that East Gwillimbury staff use on phones and tablets to make updates and edits in the field.

"This means when the data is available to East Gwillimbury, it's also available to the rest of us," Thomas said. "Historically, that would only happen monthly or if we asked East Gwillimbury staff to export it."

Maturing the Partnership

The YorkInfo Partnership coordinates data purchases for the group and has successfully driven down costs as a result of this consolidation.

"We leverage our collective purchasing power as well as each other's time and resources," Lamb said. "With 14 partners in 1, we find savings we wouldn't have found on our own."

The new seamless pattern of sharing has given partners more autonomy while also connecting each partner more closely to each other. Instead of top-down directives and guidance, the partners now serve themselves.

The collaboration also extends to knowledge sharing.

"It's really been amazing to watch our partners lean on each other and share," Lamb said. "It allows some of the smaller municipalities with just a one-person GIS shop to reach out to their 70 peers across the region who can help them be much mightier."

With the economic impact of the COVID-19 crisis, Houweling thinks greater cooperation will become more of a priority for cost savings across all local governments. "Why don't we just leverage what other people are doing? If you create a nice streetlight app, why wouldn't you put that in the Data Co-op so that the other municipalities can quickly grab it and use it, too?"

Advancing Data and Visualization Support

Enhanced data sharing has fostered more analysis to understand communities and tailor services to the people in each place.

"I've noticed more interest in things like walkability, how accessible services are, and how livable the community is," Thomas said. "That wasn't much of a focus a decade ago."

Although GIS is central to much of the management and planning work at York Region, the technology was widened by renaming the branch Data Analytics and Visualization Services. This recent move was accompanied by a four-year

road plan—a data and analytics strategy—for the whole organization. Price Waterhouse Cooper was brought in to help create the road map and work with the different data groups across the region.

"Focusing on analytical outcomes—that's how you get senior management buy-in," Houweling said. "Instead of focusing on millimeter accuracy for sewer pipes and other infrastructure, we focus on getting data good enough and using it. Our goal is to ensure our data is accurate, trusted, and accessible to provide the highest value in any situation."

The new branch name marks a shift toward a solutions-oriented mind-set focused on the capabilities of data analysis and visualization.

"It's so much better when the impetus for creating solutions comes from our business partners and is aligned to their needs rather than from the GIS side," Thomas said. "Otherwise, we're just GIS nerds trying to push technology. Once the business sees it and the business understands it, we're golden."

GIS has expanded organization-wide, and several new hires now support the region with advanced analytics. "Data scientists are mining data and doing all kinds of things I can barely conceptualize," Thomas said.

Tackling the first step in its four-year plan, leaders from York Region are now working to supplement their enterprise spatial data warehouse with an enterprise data and analytics platform. This system will deliver a geographic context to a wider number of municipal service providers and additional levels of government.

"We're not everywhere we need to be yet, but we've figured out the formula," Houweling said. "When Jack Dangermond gave us the President's Award at the 2020 Esri User Conference, he said, 'Only the people that have done these sorts of things—who have added the social grease—know what it takes.' I really love that because it summarizes our peoplecentric partnership."

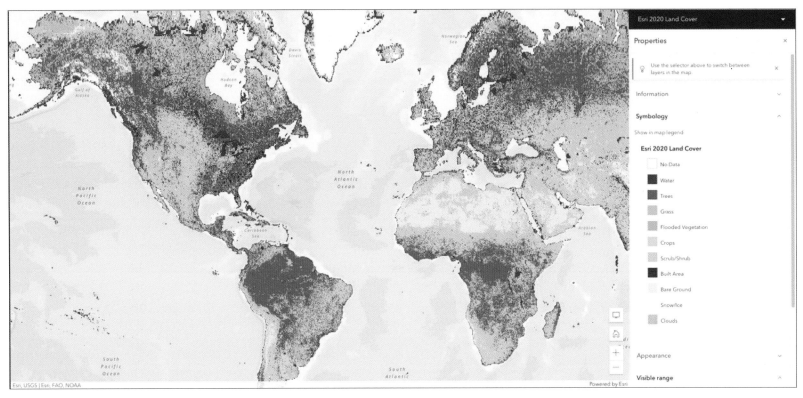

The Esri 2020 Land Cover map displays 10 distinct categories of land cover, such as trees, grass, crops, built areas, wetlands, scrub/shrub, and snow/ice.

Seeing Near Real-Time Changes on Earth with New AI Map

With commercial space travel an out-of-reach dream for most, there's an alternative for those wishing to look at the earth from space. With a click and a zoom, you can hover over any part of the planet and see it in detail, thanks to technological advances that show powerful up-to-date global images via satellite imagery.

Soon, we may be able to see global changes taking place on a weekly or daily basis.

In July 2021, partners Esri, Impact Observatory, and Microsoft released to the public a high-resolution map of the entire world as it was in 2020. With help from deep learning artificial intelligence (AI) technology, the map will soon make it possible to closely monitor in near real time the impacts of climate change and humanity's footprint.

To reach this goal, Impact Observatory used images collected by the Sentinel-2 satellites of the European Union (EU), made available through the Microsoft Planetary Computer, and trained its AI programs to build the global high-resolution land-cover map for Esri that's now free to the public.

Each pixel in the Esri 2020 Land Cover map represents a 10 by 10 m block—a significantly higher resolution than the 30 m standard—bringing the slightest detail into view much quicker. Ten distinct categories of land cover, the physical type of land, show just what's where—such as trees, grass, crops, built areas, wetlands, scrub/shrub, and snow/ice.

After being fed a heap of data and 400,000 images of Earth, Impact Observatory's AI technology quickly determined what was on the planet's surface. The data and images, previously categorized by humans, were used to train Impact Observatory for the task.

The result: a global view of land use and land cover for 2020—produced in a matter of days, instead of months. The capability should lead to answers to pressing questions about climate change and environmental crises for years to come. Everything from conservation planning to food security to hydrologic modeling can be observed.

Watching Changes as They Occur

Although the opportunity to view the planet from the outer reaches of our atmosphere has offered affordable yet grounded explorations, land-cover and land-use maps have been a necessary tool for scientists and governments since the inception of GIS technology. At the start, Roger Tomlinson in Canada invented the first GIS to inventory land for purposes of development, particularly for agriculture.

More recently, the objective has gone beyond collecting inventories to documenting climate-related changes and impactful human activity, including deforestation and reforestation, to balance land use for the better health of our planet. Consider, for example, the time-lapse feature offered by the World Imagery Wayback app, which provides more than 40 years of satellite imagery to show land feature changes over time.

Population changes, infrastructure development, and water usage—which can create environmental pressure—can be paired with land-cover conditions to facilitate understanding of environmental threats and lead to the proposal of better alternatives, wrote K. Bruce Jones of the US Geological Survey (USGS) in a paper titled the "Importance of Land Cover and Biophysical Data in Landscape-Based Environmental Assessments."

Land-cover analysis also has been used to help establish a population's risk to wildfire, flooding, tsunamis, and earthquakes, and even to assess vulnerability to contracting Lyme disease, proneness to pollutants, or the likelihood of famine. It can be used to predict crop yields and place priorities on areas to preserve.

The US Environmental Protection Agency's Regional Vulnerability Assessment Program, the National Park Service's Watershed Condition Assessment Program, and the US Forest Service's Forests on the Edge Assessment are among those using land-cover data on GIS maps, paired with other information, to aid decision-making.

Land cover can be a basis for scientific modeling, for anything from habitat suitability to environmental justice issues. A higher-resolution view and quicker turnaround time will be even more valuable to decision-makers observing ever-increasing environmental changes.

Take efforts to reforest areas of Burkina Faso in West Africa or the $8 billion continent-wide initiative to build a Great Green Wall of trees stretching across Africa. Progress can be observed at any time, from anywhere, with the Esri 2020 Land Cover map because of its level of detail and timeliness. The same could be possible for areas such as the Chesapeake Bay watershed that spans 206 counties across six northeastern US states.

In this same way, environmental leaders have relied heavily on high-resolution land-cover maps as they've monitored pollution levels for more than a decade.

When asked recently for examples of technology that have the potential to curb harmful environmental practices in their tracks, Esri chief scientist Dawn Wright pointed to the technological advancements behind the new land-cover map.

"We're providing the world with a trusted, accessible, and high resolution source for global land cover at a 10 m resolution and making this land cover on demand so we can catch things in the act, such as if there's clear-cutting going on or forced migration or illegal dumping or just for monitoring change in conservation areas," she said.

Monitoring Human Impact

Seeing where humans have spread across the landscape, in sometimes surprising locations, is perhaps the easiest impact to pinpoint, as the map comes equipped with a stark red color attached to anything built on the land.

A tiny, unfinished housing development near an Arizona ghost town outside Las Vegas may have gone unnoticed, for example, if it weren't for a handful of tiny red dots indicating structures built in the vast, untouched desert. There, the small development sits, less than an hour's drive south of the Hoover Dam—a landmark that has increasingly seen lower and lower waterlines manifesting climate change by leaving "bathtub" rings where the water level once reached. Zoomed in, looking at a basemap imagery layer beneath the land-cover data, you can see the smallest individual house. Otherwise, it would be hard to tell that anything was there but scrub.

In and around Chicago, Illinois, you can see nearly all 10 land-cover categories. Difficult to miss are the built-up areas of the metropolis that crowd the shore of Lake Michigan. The scene, highlighted by the *New York Times* in a July 7, 2021, story headlined "A Battle Between a Great City and a Great Lake," illustrates the city's vulnerabilities to climate change.

Louisiana's coastal deltas, whose stark changes prompted NASA to launch the Delta-X program, can be observed, including a shrunken Terrebonne Basin with the growing Atchafalaya Basin to its left. The increasing rate of change may become even more apparent as updates to the map are made on a more frequent basis.

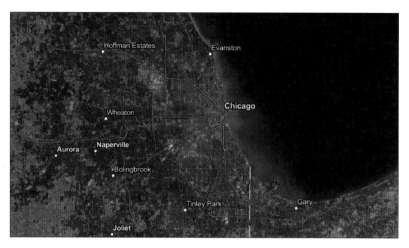
Nearly all 10 land-cover areas are visible around Lake Michigan.

The views are all part of a visual story that may have normally taken years to piece together if a team of humans had instead combed through images to identify features on the earth's surface. The work took Impact Observatory's computers less than a week to complete.

Evolution of Land Cover Maps

Until July 23, 1972, there was no way to view the entirety of the planet's terrain with the exception of photos of the earth taken from space missions that captured only glimpses. That's when NASA, prompted seven years earlier by then USGS director William Pecora, launched the first of a series of satellites to collect images of the planet. The move gave scientists a bigger, better picture of the earth's surface from above.

The first Landsat satellite (originally called the Earth Resources Technology Satellite) was equipped with remote sensors and could capture previously unseen images while orbiting the planet. It was the brainchild of Virginia Tower Norwood, a Massachusetts Institute of Technology alumnus who, while at Hughes Aircraft, invented the multispectral scanner that was integral to capturing the images for NASA and USGS. Data collected by her scanner was sent to ground stations, where it was decoded into images at a resolution that was surprisingly sharp considering the technology available at the time.

Traces of an unfinished housing project are visible in the form of tiny red dots in an otherwise barren desert area outside Hoover Dam.

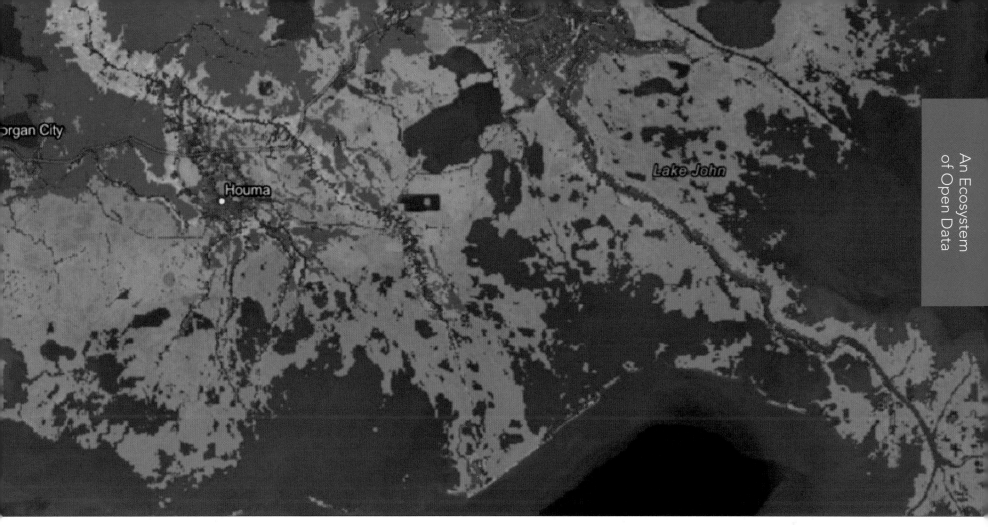

Louisiana's shrunken Terrebonne Basin *(on the right)* and Atchafalaya Basin *(on the left)* are mostly classified as wetlands, with tendrils of human settlement.

The eighth Landsat satellite was sent into space in 2013. The ninth incarnation made its way into orbit in September 2021. The satellites circle the globe, constantly collecting images, and taking 16 days to collect information about the entire planet.

Collected Landsat images have been freely available to the public since 2009. The same spirit of openness has governed images captured by the EU's Sentinel-2 satellites, part of the European Commission's 20-year-old Copernicus program to explore the earth's surface, oceans, and atmosphere. The Sentinel-2 satellites take just five days to collect imagery covering the entire globe.

The dramatically reduced time to produce a detailed, up-to-date land-cover map of the planet—between the time imagery is collected to when computers using AI use that data to categorize the land—is what carries the biggest potential for observations of the earth.

Jack Dangermond, founder and president of Esri, recently reminded the worldwide community of GIS users of the need to act quickly on climate change

"Move rapidly and urgently, because our world really needs it, because we need to create sustainable solutions," he said during opening remarks at the virtual Esri User Conference in July 2021. "It's late in the day, but it's not dark yet."

With AI and deep learning, global data can be quickly created to meet the urgency of the moment. The brightest minds can turn to land-cover maps to observe environmental statuses and trends, conduct impact analysis, assess vulnerability and risk, and run planning scenarios that focus on improving human and planetary health. Scientists and policy makers can now focus on problems and solutions rather than the process of classifying and categorizing masses of information.

Climate Action

Climate change incidents and disruptions do not exist in a vacuum. The environmental crises come in many forms and are often linked, from declining air quality, to rising sea levels, to more extreme storms, to shifting populations, to the collapse of biodiversity. Issues in one area can quickly put a strain on another, which results in compounding impacts.

If there is anything we've learned during the COVID-19 pandemic, it's that people and events around the world are tightly connected, and even interconnected.

GIS and location intelligence can help manage the complexity of the situation, from understanding weather trends and incidents to offering accurate predictions and assessments of further effects. GIS users are building solutions that factor in the analysis of risk to craft strategies and targeted responses where interventions can make the most difference.

Although climate change is global, climate-related events are local. In every area, GIS users across all domains are working on strategies to harden against the impacts. The local expertise is rolled up into regional and even national strategies. As a result, individuals and local organizations can have an outsize impact on problems as their solutions and results are replicated elsewhere.

GIS can be used to look at historical knowledge and build solutions based on what has worked in the past. Over time, the best types of projects that consider the physical conditions in each place surface and can be used by others.

In San Francisco, the San Francisco Estuary Institute has created an *Adaptation Atlas* to address sea level rise with strategies tailored for each of the many land types that will be impacted. Similar strategies are at play in Cape Cod and other areas that have seen the impacts of rising seas. As sea levels continue to rise, other localities can learn from the tools and strategies deployed there.

GIS also helps explain the geology and geomorphology of a place, as well as other earth processes and infrastructure impacts. Data science can be used to understand these processes with the analysis of large volumes of remotely sensed data, such as images and readings from tide gauges. This knowledge is then modeled against sea level rise projections to arrive at adaptation strategies.

Modeling the processes helps communities improve climate resilience while maintaining functional ecological and hydrologic systems.

One strategy that's increasingly on the table is managed retreat, a practice that some cities have adopted to buy houses and change zoning to discourage people from living in areas that will soon be under water.

Such anticipatory action is ambitious because it's a completely different way of operating that requires a shift in mind-set. Acting early can be transformative, saving lives and money, but it's not how agencies and governments normally approach pending emergencies. However, this soon could change, considering international policies that are in the works and a changing perception.

UN Secretary-General António Guterres told a humanitarian aid audience at the UN General Assembly in a September address that anticipatory action is core to his prevention agenda. And German Foreign Minister Heiko Maas promised to raise his government's funding to 100 million euros by 2023, dedicating 5 percent of its overall humanitarian funding to anticipatory approaches.

Climate change will continue to cause unpredictable events that can inconvenience or displace populations across the nation and the world. Instead of working to fix the problems of the past, it's time to shift to preparing ourselves for the future.

> *"Our life's work here at Esri is to balance human-made systems with the natural world. It's the reason we make GIS technology and support our customers and partners in applying GIS to solve the world's problems. We recognize the seriousness of the climate crisis, and we know full well that technology will be a crucial part of the solutions."*
> **Jack Dangermond,** Esri President

For Global Leaders Focused on Climate Action, GIS Technology Is Foundational

By Jack Dangermond

New commitments by 40 world leaders and more than 300 businesses to address climate change, with ambitious goals for emission reductions, make me excited about our ability to achieve meaningful change in my lifetime. In the wake of the coronavirus pandemic, our collective vulnerability helps us realize we are in this together. The pandemic disruption has provided a pause for reflection, and leaders are wise to use this time to focus on the causes of increasing climate calamities and biodiversity decline. We can, and must, take global action.

I'm heartened most by the "Build Back Better" message of the Biden administration as well as all global leaders around the world with similar sentiments, because it turns the moment into one of opportunity and level setting to ensure that everyone benefits. The GIS technology we build at Esri provides a new level of contextual awareness that gained global attention during the COVID-19 crisis. The Johns Hopkins COVID-19 dashboard was built with GIS technology. Thousands of other organizations around the world have used GIS from Esri and ArcGIS software to analyze health and safety information, understand the needs of communities, and take a data-driven approach to the pandemic's many challenges.

My life's work has focused on building decision support systems rooted in geography to combine disparate data and help people see the whole of a problem. My wife Laura and I started Esri with the enduring belief that a geographic approach can improve the natural world, our built environments, and the lives of people. I feel thankful that millions of GIS users around the world provide proof of this premise.

A Shared Perspective on Challenges

Humans continue to reshape the natural world and introduce new technologies made to improve our lives. We extract materials and burn them for energy, often to the detriment of our planet. Although we have dramatically propelled our standard of living, fueled global economies, and increased human lifespans, we now see some practices as too intensive and inherently unsustainable. With species dying around us, and factors such as sea level rise forcing us to adapt, we must retool to take on the climate crisis.

Geography, the science of our world, is experiencing a renaissance because of the nature of the problems we are facing. Maps have always been the way we understand the broader context of a plan, a place, or a situation. Today, entirely new kinds of maps and data visualizations are possible considering the integration of sensors and the data we all generate with our devices. Around the world, organizations are applying apps, drones, artificial intelligence (AI), machine learning, and the Internet of Things (IoT) to populate a real-time view of assets and people on a shared map.

Modern GIS software brings clarity to the complex climate challenges we confront by helping make sense of massive volumes and varieties of data. Datasets can be enriched, analyzed, and overlaid in ways that reveal patterns, trends, and relationships that would otherwise remain hidden. GIS shows us an accurate picture of current conditions and empowers us to model and simulate intervention strategies. We can iterate with GIS to see what decisions to make, and when and where to make them. We have seen this in action as leaders choreographed their pandemic response guided by spatial analysis and real-time awareness from GIS.

Achieving a New Resilience

The global commitment to dramatically cut back on carbon emissions will require new approaches and a much more efficient use of water, food, energy, shelter, and mobility. For humanity, this is uncharted territory.

I've had the privilege of working alongside such global conservation leaders as E. O. Wilson, Jane Goodall, and Peter Raven. These great minds have all used GIS in their work because it helps clarify their questions and guide sustainable decisions. For years, GIS has helped organizations of all sizes tackle complex problems. Now, we have the opportunity of helping them see their work in a new light, applying the ethics of conservation for careful resource use, allocation, and protection.

From city planning to asset management, our users have relied on GIS to keep their organizations strong and resilient through many crises. As the president of the United States and other world leaders call for a clear plan and real results around the climate crisis, Esri will continue partnering with its users to help fulfill their sustainability goals. Together, we will promote a resiliency that transcends the needs of any single organization or moment in time. And I have no doubt the collaborative nature of GIS work will foster new, important connections among and within organizations.

Recognizing the value of GIS at a global scale, the UN has organized a set of sustainable development goals (SDGs) with GIS as a foundation. Each participating country is collecting relevant data dealing with society, the environment, and the economy to tackle SDGs related to such challenges as water quality, forest loss, and affordable clean energy.

GIS fosters a ground truth, a common operational picture for everyone involved in solving a problem. In the context of climate change, we can use GIS to confront the consequences of centuries of shared actions. Now is a time for organizations and leaders to be bold, to face the environmental harm of our globalized economies. And that's the good news of the recent turn of events—it's a genuinely promising time for government and business, not only in terms of technology, but in terms of attitude.

How an Atlas of San Francisco Bay Is Helping Deal with Sea Level Rise

Recent flooding along one of America's great estuaries—the San Francisco Bay—is prompting local groups to act against further climate risk.

The San Francisco Bay Trail provides more than 350 mi. of walking and cycling paths through the 47 cities and nine counties that ring the San Francisco Bay, and acts as a popular viewing platform of tide cycles and abundant wildlife. Because it sits at the wetland-urban interface, the trail is vulnerable to sea level rise.

"In the San Francisco Bay, we have a long history of wetlands and ecosystem restoration for habitat conservation," said Julie Beagle, former senior scientist in the Resilient Landscapes Program at the San Francisco Estuary Institute (SFEI).

To understand the Bay Trail's vulnerability and determine ways to mitigate flood damage, planners with the East Bay Regional Park District are using SFEI's *San Francisco Bay Shoreline Adaptation Atlas*, created by Beagle and her colleagues. The atlas serves as a science-based framework for developing adaptation strategies specific to the area's diverse shoreline and that take advantage of natural processes.

"We started to get calls from regulatory agencies dealing with proposals for horizontal levees and beaches—all kinds of new living-shoreline ideas," Beagle said. "The regulatory environment, which protects the bay, has historically prohibited the use of sediment to fill in the bay, and there was no science or regulatory guidance on how to soften the shoreline to deal with sea level rise."

The standard method has been to harden the shoreline using riprap and concrete seawalls. Beagle and her team at SFEI set out to share methods and introduce green infrastructure into the region's shoreline management practices.

Starting with Location Awareness

For much of its 40-year history, SFEI has used a GIS to compile information about the estuary's chemical, physical, and biological health. GIS also guides SFEI's mitigation strategies and actions.

Wetland loss continues to be a priority for the San Francisco Bay because an estimated 80 percent of marshes and mud flats no longer exist, according to the California Natural Resources Agency. This loss eliminated wildlife habitat and storm-buffering capacity, which will be vital as storm intensity increases and seas rise. In 2016, voters in the nine Bay Area counties approved a bond measure to provide $500 million over 20 years for wetland restoration. In some areas of the bay, restoration will suffice, but others will require hard infrastructure, such as armoring the shoreline and raising levees.

"There are a lot of big questions, and we used GIS to look at the historical knowledge about the bay to determine what makes sense geomorphically," Beagle said. "After Hurricane Sandy, there were a lot of images of marshes being built at the bottom of Manhattan when we know there's really deep water there with high wave energy—that's not going to work, and it won't last. We spent five years on the atlas to show the best types of projects for the physical conditions in each place."

30x30 Sets Conservation and Biodiversity Goals

A growing number of governments have pledged to conserve 30 percent of land and coastal waters by 2030 to address ecosystem degradation and mass extinction.

In California, Governor Gavin Newsom signed an executive order in October 2020, enlisting the state's vast network of natural and working lands—forests, rangelands, farms, wetlands, coast, deserts, and urban green spaces—in the fight against climate change.

GIS is a core technology in this evidence-based plan, which includes the formation of the California Biodiversity Collaborative and an upcoming Natural and Working Lands Climate Smart Strategy to guide climate action in California.

This effort will also create a GIS called CA Nature to integrate data, build a shared understanding of the natural world, establish a baseline to gauge 30×30 progress, and identify conservation opportunities. Plans will honor the sovereignty of tribal nations and aim to ensure equitable distribution of nature's benefits to communities of color and low-income communities.

Beagle also hopes that use of the *Adaptation Atlas* will help instill a regional approach spanning all shoreline jurisdictions, including cities, counties, wastewater treatment plants, highways, and private landowners.

"It's about collaborative planning and collaborative science for resilience," Beagle said. "If each entity is just going to plan for its own piece of shoreline, it's not going to be effective, it won't be resilient, and it doesn't have any ecological benefit. The poorest communities get left behind, and the richest communities pay to build themselves out of sea level rise. That's what we see happening right now."

In addition to the equity issues Beagle raises, there is also the simple fact that nature does not operate according to jurisdictional boundaries or the built environment.

"The idea is to capture the processes of watersheds, creeks, and the shoreline to plan holistically," Beagle said. "Another way to think about it is nature's jurisdictions."

The San Francisco Bay, one of America's great estuaries, is undergoing restoration.

Imagery Deepens Understanding

SFEI collects many layers of information about the bay, estuary, and watersheds, and it acts as a state spatial data center. All bay restoration projects are entered into SFEI's EcoAtlas to be tracked over time by regulatory agencies monitoring compliance. SFEI purchases aerial and lidar imagery to monitor change and accurately measure elevations. For more thorough or immediate imagery analysis, the institute deploys drones to key areas of interest.

"With the drone, we're able to survey right after a storm or return at the right time of year to see what's changing," said Pete Kauhanen, GIS manager and drone expert at SFEI.

The flexibility of the drone platform allows SFEI to monitor sediment volumes in stream channels and tidal marshes, monitor restoration sites, and capture harmful algal blooms. SFEI also uses drone imagery and a sophisticated machine learning algorithm to count and monitor trash.

Many of the drone missions fulfill multiple objectives.

"Flying over the Corte Madera Marsh allows us to see patterns of erosion or progradation of the marsh edge, and at the same time we can see the state of trash in the marsh," said Tony Hale, program director of the Environmental Informatics Program at SFEI. "We're able to look at both the built environment and the natural environment and will continue to look back at the record to see what has changed."

SFEI's drone team also assisted with the institute's Resilient Landscapes Program to quantifying the impact of winter storms. High-resolution drone imagery and tools that automatically calculate volume provided an accurate measure of sediment and its movement. The Resilient Landscapes team analyzed the imagery to identify erosion and the surge channels that have eaten away the marsh. The work also captured an area with a degrading levee that caused the beneficial expansion of a mud flat.

By using imagery to observe and analyze natural processes, SFEI teams can guide strategy and measure the performance of different management practices.

"Imagery also helps us gauge the amount of sediment that we need to supply the marshes so they can keep up with climate change," Hale said.

Accommodating Differences

The Bay Area is home to a broad diversity of people, topography, tidal patterns, and microclimates. Strategies that work in one city for one stretch of shoreline may not work in a neighboring city. The Adaptation Atlas took the approach of cataloging and describing the 30 distinct operational landscape units (OLUs) of San Francisco Bay's 400-mi. shoreline.

"We look at what makes sense for the land use that's existing and for what could be there in the future," Beagle said. "The Adaptation Atlas gives us a framework within which we can work using operational landscape units that are based on our understanding of the underlying geology and geomorphology."

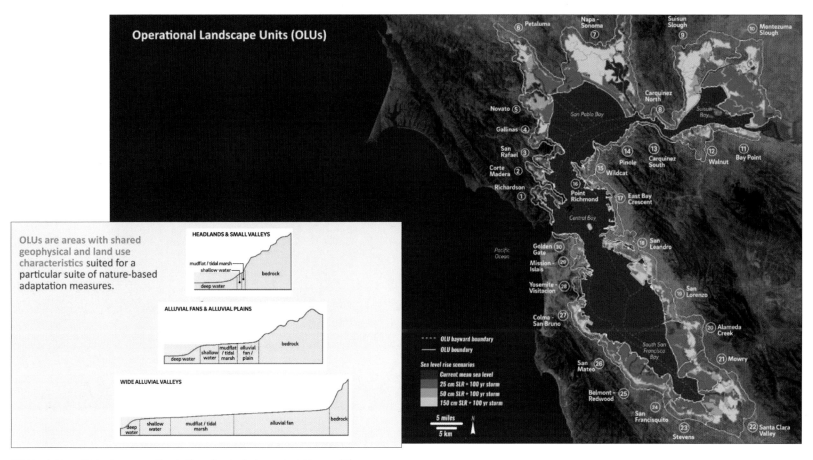

OLUs define distinct areas with similar physical characteristics and the map shows how the different OLUs ring the bay alongside expected sea level rise.

Modeling the processes at play helps SFEI tackle the bigger question of how to improve climate resilience while maintaining functional ecological and hydrologic systems.

"By using big spatial metrics, we work to understand how high and wide certain features need to be to knock down waves in each location," Beagle said. That knowledge is then modeled against sea level rise projections to devise adaptation strategies.

From Cleaner to Climate Ready

From the 1850s to the 1960s, the San Francisco Bay was often filled in to create land for housing and infrastructure. A lot of homes sit on land that's built from artificial fill, which poses its own problems in this earthquake-prone area. Wreckage from the catastrophic 1906 earthquake was systematically dumped into the bay and surrounding marshland. Some homes even sit on former garbage dumps as there have been many open landfills surrounding the bay.

"They used to talk about how bad the bay smelled," Beagle said. "Then with the environmental movement and changes in laws, you could no longer fill the bay, and we began cleaning it up."

Although nobody can yet claim that the bay is clean, years of remediation have improved water quality and restored vital habitat.

"Now, it's less about fixing the problems of the past and more about how we prepare ourselves for the future," Beagle said. "We don't want to wind up with seawalls everywhere. It's about how to have a shoreline that's still ecologically functioning, accessible to people, and equitable."

Uppsala, Sweden, Creates a Detailed Digital Twin to Enhance Sustainability

Sweden, like many other countries, struggles with a housing shortage and lack of affordability. This isn't the first time the country has faced such a crisis. In the 1960s, the Swedish Parliament passed the Million Dwellings Program, which created a million new homes between 1965 and 1974 to accommodate the wave of baby boomers reaching adulthood. A similarly urgent movement to add housing is now under way.

City planners in Uppsala, the country's fourth-largest city, are doing their part to design a new district with 33,000 new housing units to accommodate 50,000 new residents by 2050. Today, the municipality of Uppsala has 230,000 residents and is expected to reach 350,000 by 2050, which is more than a 50 percent increase in the next 30 years. Much of this expansion will occur according to Uppsala's comprehensive plan, but an

area equivalent to 250 city blocks is being designed around green-growth concepts.

"Normally, we make plans for a couple of hundred new housing units here and there," said Svante Guterstam, strategic community planner, City of Uppsala. "Now, we're planning an entirely new district."

Acting on a Mandate to Innovate

As Sweden's fastest-growing city, Uppsala attracts new residents based on its reputation as a research center and its ambitious sustainability policy. For instance, Uppsala University—founded in 1477 as the first university in Sweden—has long focused on the natural sciences, its motto containing the phrase "truth through nature". The Swedish University of Agricultural Sciences, also located in Uppsala, recently created the Uppsala Green Innovation Park to incubate businesses focused on sustainability and partnered with the city to research urban development projects.

The city's green approach became world leading over the past decade with a pledge to be free of fossil fuel by 2030 and be zero carbon by 2050. The Uppsala Climate Protocol—acting as a local Paris Agreement—brings together 42 local businesses, public agencies, associations, and the city's two universities to meet its climate challenge. Many of these same stakeholders are involved in creating the Uppsala Climate Roadmap, which will create a guide on how to phase out fossil fuels.

With this background, it's no surprise that when city planners set out to create a southeastern city district, they adopted the data-driven and modelcentric approach of geodesign powered by a GIS.

"We're testing out cutting-edge techniques and working to find new solutions for becoming more sustainable," Guterstam said. "With two universities in Uppsala, we have a close connection with the academic community where we can explore creative and innovative approaches that include new actors and ways to have a smaller climate footprint."

Uppsala planners are concentrating on a sustainable urban model that adds to residents' quality of life, doesn't subtract from biodiversity or degrade the environment, and cuts carbon emissions. A detailed zoning plan and 3D model built with ArcGIS® Urban software help the planners visualize and present plans for the new city district.

"We value the model a lot during the working process because we can test different outcomes and see—and also show, for example, the politicians—that if we do it this way, this is the outcome, and if we do it that way, it's another outcome," said Marie Nenzén, strategic community planner, City of Uppsala. "With the model, the outcomes become very clear and visible."

Reinforcing Uppsala's Green Identity

The Swedish government promotes the planning of districts and even new cities to stimulate more housing. In return for Uppsala's commitment to growth, the state will invest in new railway tracks to Stockholm, a new station, and a tramway.

The new district will be built on forest land currently owned by the city. Many of the trees will stay within green corridors, and the blocks have been designed so that large groupings of trees can occupy courtyards with houses built around them. There's also an emphasis on leaving habitats intact and taking care to maintain natural systems such as groundwater flow.

"The natural areas that are close to the city, and sometimes in the city, are very dear to the people of Uppsala," Nenzén said. "We want to leave as much as possible undeveloped and untouched while also addressing the demand for housing. It's a constant balancing act of trade-offs."

The city and its partners are also exploring how to update traditional city infrastructure—such as water, wastewater, transport, and energy systems—to be more natural and energy efficient. "We need to find completely new ways of building this infrastructure to have circular systems," Guterstam said.

The new city district will create sustainable living as well as a new city center south of the historic downtown.

"When fully built, it will be a dramatic change compared to the current city that will have an impact on the identity and concept of Uppsala," Guterstam said. "There will be a different balance with another urban center and southern entry."

Given this transformation and the need to clearly communicate information, the detailed planning proposal includes an interactive model that allows anyone to view what has been planned.

"The interactive 3D model gives you a feeling and impression of this future district in a way that written documents or 2D maps can't," Guterstam said. "Now we can zoom and fly around this new urban area and get quite a good sense of what is coming."

Modeling to Conceptualize and Consider

Early in the planning process, the planning team struggled with the amount of tabular data about project requirements. The number and size of spreadsheets kept increasing, and data in that format made it tough to engage and collaborate with the multiple city departments involved. It was also hard to conceptualize and share a physical representation of the complex district master plan.

"We started off thinking of the model as a tool for communication," Guterstam said. "After we built the model, we realized that it provided a way to study and share scenarios in a simple way that makes urban planning much more understandable."

The scope of the southeastern city district goes beyond past plans and exceeds the work experience of most of the project's planners, architects, engineers, and researchers. The model helps everyone see the project's scale and conceptualize the urban form themselves to grasp what works and what doesn't.

"It's a learning process to understand the kind of density that will fit within the area of the agreement," Guterstam said, stating that "3D modeling is very useful to try different building and block sizes and what scale it produces. Does it resemble Manhattan or a historic European city?"

The 3D model is built on top of a digital elevation model captured with lidar scanners to show the real topography. Lidar also captured existing buildings and all the trees. The realistic model enables everyone to see how the proposed development will look in the context of the current city. The model has helped the planning team test approaches and configure and test new areas, and ultimately it will help the city and its residents make informed and sustainable decisions.

Uppsala leaders are considering a range of factors to make the city more sustainable, including mandating wood building materials since the country is richly forested and wood is a renewable resource. They are also looking closely at mandates on energy, water, and resource consumption.

"With this smart model, we can also get indicators for how sustainable the plan is," Guterstam said. "In order to achieve the targets for 2030 and 2050, we need to back-cast to see if what we plan will help us reach our ambitious climate and sustainability goals. We have a very good chance of influencing

these kinds of decisions now in the planning process, because once the district is built, it's hard and expensive to change the urban systems to enable and encourage sustainable lifestyles. We need to identify solutions and techniques that actually take us where we need to go."

Residents of all ages enjoy the proximity to nature in Uppsala.

Orchestrating Denser Expansion

The rail network and new tramway form the spine of the new city district, matching a growing pattern of urban planning called transit-oriented development that reinforces the value of denser urban pedestrian-oriented areas centered around public transit and having strong cycling connections.

Even before the southeastern city district was conceived, transportation was a strong focus of Uppsala's efforts to become carbon neutral. Uppsala's buses are powered by renewable energy and biofuel produced from local food waste. New bike lanes and a two-story bike-parking garage next to the train station encourage people to leave their cars at home. The city also launched a Climate Made Easy campaign in 2019 to help residents reduce the carbon emissions they are personally responsible for.

"With coherent urban planning we can set incentives for sustainable lifestyles where you're not car-dependent and can live without a car by using public transit," Guterstam said.

Accelerating Input to Spur Advancement

Though planners took care to reveal the right level of detail in the model for the early planning phase, they anticipated confusion from residents who weren't used to seeing such detailed plans.

"In public communication, it can be challenging for many to

understand that the plan is very much a work in progress still," Nenzén said. "We get questions that we can't answer at this stage. But it also shows that the model spurs a lot of curiosity, and the public wants to see and use this model."

In fact, of all the elements that Uppsala has uploaded to the 2050 planning site, the model has received the most web traffic. The model has also become popular with the many stakeholders working on the plan.

"It was fun to see how people on other teams got interested in the model," Guterstam said. "They were asking about how we're using it and if it could also be applied to their projects. It's different from how we've worked in the past. We have full control of the model instead of having an external consultant architect create a model for us and who we then have to ask for changes that we sometimes would get back weeks later."

For city staff, the ability to quickly make changes and see the impact of alternative designs is helping drive adoption of new workflows and the use of new tools.

"Some have questioned the flexibility of the plan," Nenzén said. "Demonstrating how easy it is to make changes in the model makes it easy for people to understand that we can change it when it needs change. For instance, when the demand for more freestanding houses was voiced, we could use the 3D model to identify locations and study how it would impact the height and density of the remaining apartment blocks."

The accelerated feedback loop fits the moment because Uppsala's agreement to expand comes with a timeline. The plan must be in place, and the city has to start building 800 new housing units per year, by 2025 to secure the national government's infrastructure investments. Before the city begins building, a lot of parallel processes need to take place.

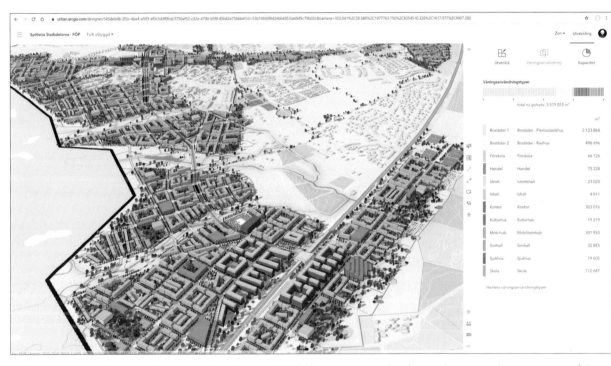

An interactive massing model presents the planner's initial plans to convey the density, layout, and environment of the southeastern city district—helping stakeholders see as planners and understand what they see.

Planners and architects are conducting public consultations according to planning law. The 3D model is being refined to visualize streetscapes, determine the scale of the buildings, consider the right mix of forest and trees, and review water and stormwater processes. The visualization will also help determine the materials used to construct the buildings and infrastructure.

"During this process, it has become clear that this tool has a lot of potential as we get more detailed and have more data," Nenzén said. "For example, we want to test and see the outcome of daylight and shadow studies and make more connections to sustainability planning."

Urban Parks Play a Big Role in Curbing Inequity and Climate Impacts

As the locus of most economic activity and greenhouse gas emissions, cities around the world face the escalating challenge of understanding and managing climate change. Cities are home to half the world's population, and as urbanization increases, so too will climate risk. This makes urban strategies to curb emissions even more critical. Many of these strategies are being built around location-specific insight from data-driven maps and spatial analysis using a GIS.

Although cities stand out as primary contributors to climate issues, they are also the biggest victims of climate change. Most lie near coastlines or inland waters, putting them in immediate danger from rising sea levels and extreme weather events.

Taking a more granular view, climate vulnerability across cities is unevenly distributed. Low-income communities suffer more than others, often for deeply rooted historical reasons. As the crisis intensifies, these urban climate inequities will, too.

City Smarts and Smart Cities

Leaders in many major US cities are approaching climate risk and social equity issues by rethinking public parks, guided by GIS maps. To that end, The Trust for Public Land (TPL)—a San Francisco-based nonprofit that advocates for the creation of parks and preservation of green spaces—has partnered with governments in Boston, Los Angeles, New York, Denver, and other US cities. Ten years ago, TPL launched its Climate-Smart Cities™ program, which addresses unique climate-related problems in urban areas.

"Parks promote health and improve well-being, build social cohesion when communities come together outdoors, and make cities more resilient to climate change," said Lara Miller, TPL's senior GIS project manager.

TPL staff use GIS maps, for instance, to identify which communities are underserved by parks or identify how many people within a city can reach a park within a 10-minute walk.

Through collaborative work with local governments, TPL helps cities build green infrastructure with an emphasis on climate change resilience. Projects might include new trails and transit lines, tree canopy to increase shade and reduce heat islands, parkland and playgrounds that also work to reduce flooding, and shoreline parks that protect coastal cities from sea level rise.

The Climate Crisis and the Crescent City

Just a few years into recovery from Hurricane Katrina, New Orleans, also known as the Crescent City, became one of the first TPL Climate-Smart Cities partnerships. Causing more than 1,800 deaths and $125 billion in damage, Katrina was a harbinger of the disruptive weather events that will increase with climate change.

As the city prioritized improving climate resilience, TPL facilitated the creation of "green schoolyards," replacing concrete surfaces prone to flooding with gardens of native plants that absorb rainfall and runoff. Several other interventions, including wetlands restoration and storm water catchment basins, also addressed flood concerns while increasing open space access for neighborhoods that did not previously have it.

The New Orleans collaboration set a Climate Smart precedent for encouraging data-driven solutions. TPL developed Climate-Smart Cities New Orleans, a GIS-based tool to plan and implement projects. Climate-Smart Cities New Orleans stores and integrates location-specific datasets and projects them as layers on a smart map. For New Orleans, the tool included flood data, public health and household income information, and a map layer of green space access across the city.

"We tailor our work based on a city's biggest challenges and goals," said Taj Schottland, The Trust for Public Land's senior climate program manager. "For a city like New Orleans, it's no surprise that the emphasis would be on flooding and absorbing runoff. In other cities, sometimes it's transportation or urban heat islands that emerge as the major priority."

These tools help communities understand risks but also guide local action. In New Orleans, "whenever someone proposes a green infrastructure project, the city requires the applicants to use our decision-support tool to justify why they want to do it in a given location," Schottland explained.

The Big Drawdown

The paradox that cities face in the age of climate change is that they are, as the UN put it, both "the cause of and the solution to" this existential crisis. Even as they compound the problem, cities embody many of the best solutions.

For example, density can promote sustainability. Research suggests that if the US were to take moderate steps toward promoting housing density and improving transit, by 2030 the country could cut its emissions by a third.

The goal of the Climate-Smart Cities program is to help cities manage the effects of climate change through targeted green infrastructure projects. TPL recently expanded its climate-related activities to help cities understand the part they can play in alleviating the root cause of the climate crisis.

Using the same type of GIS-based tools employed by Climate-Smart Cities, TPL has partnered with the Urban Drawdown Initiative—a project of the Urban Sustainability Directors Network—to advance nature-based solutions that capture and store carbon and, critically, deliver local health, equity, and economic development benefits for communities.

"We're working with a host of cities, Colorado State University, and other researchers to quantify carbon capture by urban green spaces, and then model how different nature-based interventions can increase active carbon capture," Schottland said. "This work is groundbreaking and has the potential to transform our understanding of the role regreening our cities can play in the climate crisis."

Environmental Justice

Intense media coverage of Hurricane Katrina highlighted the equity component of natural disasters—who is impacted more than others and how those affected are treated in the disaster's aftermath.

New Orleans' Black residents, faced with disproportionate poverty rates, were less likely to have a means of escaping the city. In the most symbolically stark example, evacuees were prevented by police from walking over the Mississippi River bridge that connects New Orleans with the town of Gretna.

In the ensuing years, environmental justice—the equal treatment and involvement of all people in environmental— has become more mainstream.

TPL's Climate Smart tool has been one way for cities to highlight environmental justice. In a recent collaboration, TPL staff helped Los Angeles planners identify areas of extreme heat within the city to prioritize heat reduction efforts.

"We were able to show the city the census blocks, and the data relating to the census blocks, overlaid with heat islands," Schottland said. "Low-income residents who live in hotter neighborhoods with less tree canopy are less likely to have the resources to pay for air conditioning—or they may be more likely to work outside instead of having an air-conditioned office job. So we want to direct city investments to these neighborhoods and protect those who are experiencing extreme heat."

Maintaining Equity

As our society has come to understand the need to tackle social inequity and the climate crisis, TPL staff have refined their approach. As TPL builds new GIS-based tools and creates new parks, the staff help cities guard against unintended consequences of neighborhood improvement. New green infrastructure such as parks, gardens, and playgrounds can make a neighborhood more desirable, a process researchers call "environmental gentrification."

Including Equity and Urban Access in 30×30 Conservation Plans

In January, as one of his first acts in office, President Joseph Biden signed a sweeping executive order outlining several directives related to the climate crisis. Among them was a pledge to conserve 30 percent of the country's land and waters by 2030. Several countries and the State of California have set similar goals, a movement broadly known as 30×30.

The idea conjures images of wilderness areas far from the city, pristine realms unsullied by humans. As environmental science has developed in recent years, however, experts no longer consider this an optimum way to approach conservation on a large scale.

The indigenous history of land in the US—and much of the world—reveals a more fluid boundary between nature and humanity. With that in mind, the Biden administration has pledged to work with rural officials and tribal leaders. They will aim to link disparate swathes into larger expanses, furthering the 30×30 effort while respecting the sovereignty of tribes to hunt, fish, and gather on their land.

The Center for American Progress has argued that 30×30 plans should expand access to nature. Their report, "The Nature Gap," found people of color are more likely than White people to live in nature-deprived areas, and low-income communities are more likely to experience nature deprivation.

The effort to distribute nature's benefits more equally requires geographic context. Maps layered with all relevant data can reveal inequities in access to green space, supporting 30×30 initiatives.

"There are examples of green space being created in a neighborhood and people being displaced," Miller said.

The ParkScore for New Orleans benefits from destinations like City Park, a popular gathering spot.

index quantifies how well the 100 largest US cities provide communities with park resources. Cities are awarded points based on analysis of four important characteristics of an effective park system: acreage, investment, amenities, and access. ParkScore is adding a fifth characteristic, equity, to better assess how park systems are providing equitable access to the health, climate, and community-building benefits of parks and green space.

TPL developers also designed ParkServe, another GIS tool that quantifies how well the cities meet community needs for green space. Using ParkServe, city leaders and park advocates can access TPL's comprehensive database of local parks in nearly 14,000 cities, towns, and communities to guide improvement efforts. This data underpins TPL's engagement with cities.

"During planning and project implementation, it is critical we engage community partners to understand gentrification or displacement concerns and bring those concerns into our park development process."

The Climate-Smart Cities tool provides a way, in effect, to knit together TPL's three pillars of parkland value—health, equity, and climate—into a more expansive view of the modern city.

"The program has always had an equity lens, but our thinking about what equity itself means has evolved and continues to evolve," Miller said. "When it began, we were thinking more in terms of physical access, like who is without air conditioning or those who can't walk to a park within 10 minutes of where they live. Now we can also examine other issues related to equity, like access to information and decision-making."

This evolution continues in another national tool developed by TPL, the ParkScore Index. This

Trees, here waiting to be planted, soften urban landscapes and provide much needed shade and many more ecosystem services, including filtering air and stabilizing soil.

Finding a Home for Solar: Kentucky Maps Prime Renewable Energy Sites

The state of Kentucky is undergoing a transformation from a coal powerhouse to a compelling locale for renewable energy generation. Once the leading producer of coal in the US, and still one of the top three coal-producing states, Kentucky is nonetheless looking toward a future powered by alternative energy sources—hydropower, biomass, and solar.

In the face of declining coal use and production coupled with increasing interest from corporate buyers of renewable resources, Kentucky is experiencing a significant growth in interest from developers looking for solar power project locations. With its robust infrastructure and available space, including previously used mine lands, Kentucky bills itself as an ideal choice for solar production sites. But where does it put them?

The Kentucky Energy and Environment Cabinet set out to answer that question. "For several years, as solar has come down in cost, it's become more of an option here in Kentucky," said Kenya Stump, executive director of the Kentucky Office of Energy Policy. "We had a lot of questions from people: 'It seems like our mine lands would be great for solar,' or 'I don't know why we don't put solar on our mine lands.' And to me, that was always a geospatial question. Where should new sites go?"

Until recently, there was no mechanism in place for the Office of Energy Policy to receive and respond to site inquiries, so developers often selected sites with little to no input from the state. "Solar is so new in Kentucky that we had solar developers making siting decisions, but we didn't understand why they were choosing their locations," Stump said. "We would just get notified that a new solar project was going here or going there."

Staff from the Kentucky Cabinet for Economic Development would pass along inquiries to colleagues at the Office of Energy Policy, who, initially, had no way to gauge suitability. Traditional industrial development sites weren't meant for solar, and siting characteristics such as topography, slope, or the presence of threatened species had to be considered. Teams at the Office of Energy Policy used their technology to create the Solar Site Potential in Kentucky platform, guiding solar developers to prime locations.

A Smarter Solution for Site Selection

At the start of the project, the team from Kentucky found few best practices to follow from other states. With input from the state's GIS analysts and Esri, Stump and others from the Kentucky Energy and Environment Cabinet applied GIS to conduct site suitability analysis on land parcels available for development. The analysis evaluates sites based on criteria impacting the technical feasibility of construction, and then gives each parcel a score.

Stump's team collaborated with mining consultants and solar developers to collect relevant data layers and determine the criteria needed to compare sites. Support provided by

the KY Geography Network (KYGeoNet), the geospatial data clearinghouse for the Commonwealth of Kentucky, facilitated this collaboration. Melissa Miracle, an IT consultant and project manager at the Kentucky Energy and Environment Cabinet who is deeply involved in KYGeoNet, worked closely with the state's nature preserves to include data on local endangered species. "That was a tough one because we didn't have direct access to the raw data," Miracle said. "So we worked with the folks within the agency who handle the nature preserves to create the layers that we needed. It was important for them to be involved so the developers would know where they can and cannot build." Miracle also worked with the Natural Resources division to create a better understanding of the state's mines.

"That was a whole learning curve," Stump said. "The attributes, how they code things, understanding mining reclamation terminology—all that was huge."

Through this collaboration, the team addressed the multifaceted concerns of solar plant providers—including favorable slope, land classification (barren land, mixed forest, and cultivated crops), access to electric transmission lines, population density, proximity to the habitats of threatened species, and status as federal or protected lands.

Communicating with Communities

Another key benefit of the Solar Site Potential in Kentucky platform is the transparency it offers to communities that may be affected by possible development. "Our land-use planning is done at the local level. So this tool also helps our local communities understand the reason why developers are looking at their lands—whether it's because they have the right slope or they have access to transmission or other characteristics," Stump said. "It's still up to that community to decide what they want to do with the land."

The platform also helps community stakeholders find out exactly which local land is being considered for solar projects. This information influences personal and local planning decisions, such as safeguarding land for specific uses or protecting views.

"This tool couldn't have come at a better time because it shows stakeholders that we are actively trying to find places that were not prime farmland," Stump said. "We have a layer in the tool that you can turn on to show the land-use classifications, which can also inform the developer in the conversations with the community."

Plans are also under way to add 3D enhancements to the platform to make it easier for users to visualize the areas under consideration for development. "We're working on creating 3D map scenes in some of the prime areas for solar across the state," Miracle said. "This would allow developers to 'fly in' and see the slope, the terrain, a whole new view."

New Applications, New Opportunities

Kentucky's success in analyzing solar siting potential has created opportunities for collaboration with other states.

Former mine sites hold potential for large solar energy sites.

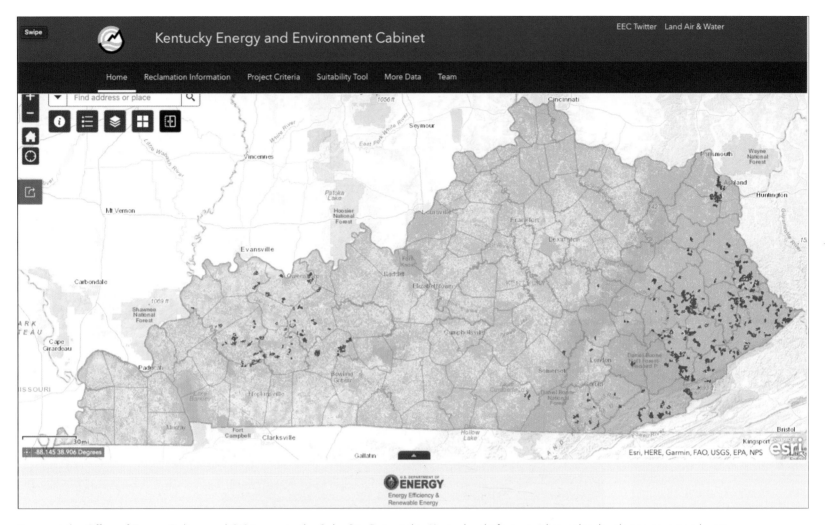

Teams at the Office of Energy Policy used GIS to create the Solar Site Potential in Kentucky platform, guiding solar developers to prime locations.

"Given the work our GIS group is doing, we're known nationwide as the office you go to if you want to learn how to get started with GIS," said Stump. "We're answering energy questions from other states that, at the heart of it, are geospatial questions. Other energy offices are beginning to see the light—that they need to know where things are going to occur, where they should occur, and where they could occur, before talking to stakeholders and discussing policies."

Kentucky's GIS experts envision using the technology to adapt the siting platform tool to incorporate environmental justice data.

"It's going to be another lens by which we look at everything we do, from emergency response to permitting to siting facilities and economic development projects," said Stump. "Right now, we're assessing where the datasets are. We know the Environmental Protection Agency (EPA) has the EJSCREEN environmental justice tool, the Census Bureau has Community Resilience Estimates, the Department of Energy has the LEAD tool to examine low-income energy affordability, but how do we bring them together? What does environmental justice for energy look like in Kentucky? That's a big where question. We're really excited about where we can go."

NEW ROAD
LAYOUT
FOR SOCIAL
DISTANCING

Infrastructure

A good deal of local government spending aimed at pandemic recovery is going toward hardening infrastructure and building in resilience in the face of growing impacts from climate change. Mitigation of climate vulnerabilities and preparedness for all types and sizes of disaster have become a central consideration for all new infrastructure projects.

A resilient system is a system that can survive shocks, especially ones that are hard to anticipate. GIS technology has played many critical roles in dealing with devastation from hurricanes and other disasters, including modeling consequences in real time to guide responders to the most urgent needs. Geospatial tools have allowed many vulnerable areas to achieve new levels of preparedness, with mitigation measures determined through GIS analysis. With the contextual awareness of GIS, communities can rebound more quickly with fewer long-term effects.

In recent years, the concept of geodesign has put a finer focus on sustainability. It's a growing practice that combines design with the more scientific task of assessing impacts. With this datacentric approach, designers and engineers implement projects to design with nature, not against it. Careful analysis yields design that best supports natural systems and enhances livability. Geodesign practitioners take advantage of the geographic knowledge contained in GIS and the detailed designs created in building information modeling (BIM) to visualize and model sustainability in multiple dimensions.

Geodesign supports urban planners, landscape architects, designers, engineers, and stakeholders with a blend of science- and value-based information for understanding the complex relationships between human-designed structures and the changing environment.

Using geospatial tools, geodesign practitioners look at census and mobile data to consider the human dimensions of a place; compile and query environmental data to quell impacts on nature; and model and run scenarios to consider the economic aspects and benefits of a project. The methodology helps assess risk, develop strategies, adapt to changes, test scenarios, and monitor the results. Geodesign also facilitates community feedback on the most suitable, environmentally friendly, and sustainable options for a space.

Geodesign is equal parts analysis, design, and iteration. At the outset, it promotes the careful consideration of a place using data and analytical tools. It facilitates design input from a cadre of disciplines working through a shared geographic perspective. And it seeks to iterate and test scenarios in the digital world to avoid making mistakes in the real world.

When using GIS to design infrastructure, the goal is to imagine the best possible purpose of a place, and to consider what the world could become. GIS places an emphasis on understanding all the consequences of development or infrastructure construction before those consequences become impacts. Geodesign factors in natural networks and impacts to ecosystems, along with renewable energy and carbon neutrality. It imagines what's best for the present and into the future, with the design of infrastructure that will support the sustainability of communities.

5D: The New Frontier for Digital Twins

When the Long Island Rail Road (LIRR) began the latest phase of renovations on Jamaica Station—the fourth-busiest rail station in North America—designers knew they needed to take a digital twin approach to project management.

A virtual 3D model of the rail line would help the interdisciplinary project team visualize the complex choreography of construction stages unfolding in a dense urban environment. The latest phase of the $1 billion Jamaica Capacity Improvements (JCI) project would involve large-scale repairs, including replacing critical power and signal systems, extending station platforms, and building two new tracks, all aimed at increasing capacity and expanding ridership through the bustling Long Island transit hub.

The infrastructure firm HNTB Corporation won the LIRR design contract in part because it put forth an innovative solution that went beyond a 3D digital twin. Using GIS technology, HNTB pioneered a 5D project management strategy that integrated interactive 3D models with the additional dimensions of time and cost.

This 5D approach empowers project leaders to track and predict how design changes might affect scheduling or construction costs and to adjust their decisions accordingly. It's a strategy that illustrates the direction that many firms across industries are beginning to take, in fields from supply chain management to ski resorts. Pairing the insights of GIS—known as location intelligence—with digital twin models, business leaders can combine easy-to-understand visuals with a powerful geospatial engine for data analysis.

Press Play to View Your Project in 5D

With HNTB now nearing the finish of the design submittal stage, the design team is using the 5D digital twin model as it prepares for the construction phase. While looking at a virtual replica of Jamaica Station tracks, project planners can pull a slider along the bottom of the screen and scroll through 36 months of forecast progress, seeing how far a new section of rail will extend by March 2022, for example.

Alternatively, an executive can press Play on a three-month period of construction and watch a split screen that pairs an animation of a pier being built with a graph that charts the costs tied to that portion of the project.

The ability of GIS to integrate different types of data and software in a cloud-native environment makes it possible for anyone associated with the LIRR project to access the visualizations via a web browser.

"A solution like this was critical because trying to convey these complex designs and this complex phase of construction is always a challenge for all the stakeholders," said Jeff Siegel, a vice president and the Technology Solutions Center director at HNTB. "Now, because of the way we've done this, they can pull up and play what the actual construction phasing will be and what the outlay of cash will be based on estimates down to the unit level."

A Web-First Philosophy of GIS Fuels 5D Interactivity

Jamaica Station is part of a heavily trafficked transit corridor that transports over 200,000 passengers a day. The redesign project was important to many stakeholders, including partner agencies, local politicians, and transit officials. A 5D digital twin that was opaque or designed only for those with engineering backgrounds wouldn't work—city council members might not be knowledgeable in AutoCAD.

After Siegel and the HNTB team surveyed the options, they concluded that GIS was the best foundation for executing a 5D digital twin plan of the JCI project.

The 5D digital twin created for the project includes details on the rail line's connections to the community.

"GIS is really the only platform that can truly provide that sophisticated integration or convergence of both the location of something as well as the attributes," Siegel said. "That enables such a strong coupling of the analytics when you want to pull all of that together."

The importance of communication and collaboration on JCI led Siegel and Darin Welch, the project lead and associate vice president of geospatial and virtual engagement solutions, to prioritize a cloud-based, web-first approach. In the past, shareable maps were sometimes among the last things to be produced as a project was wrapping up. That set the stage for a new era, with GIS as a key part of digital innovation delivery.

Welch, Siegel, and their team of technologists established a web-based GIS strategy to drive the 5D project from the start, acting as a centralized, authoritative data source where information would be updated continuously and open to any team member—internal or external—who needed access. With that model in place, workers who went into the field to perform inspections of the track and take photos were able to upload such data directly into the GIS platform, making it available within a 5D view.

"It's really changed the way many people in our organization think of GIS, because location is such a critical dimension of the data we leverage within the AEC space, and it's an approach we continued perfecting on Jamaica Station," Welch says. "We see much more openness and willingness to leverage the power of GIS because we can take, with confidence, design information overlaid with other rich GIS datasets and tapestry, then allow that to influence our decision-makers, who are sometimes public."

Geospatial Problem-Solvers

It was this proclivity to pursue new ideas that landed Siegel and Welch the assignment to implement the 5D digital twin approach. The original idea for the 5D digital twin came from HNTB's New York-based design team to help LIRR manage the complex schedule and budget of the Jamaica Station project.

The 3D animators on the HNTB marketing team created an animation for the company's pitch to LIRR decision-makers. Once HNTB won the project, it was up to Siegel, Welch, and their colleagues to turn the concept into reality.

Siegel and Welch are part of the Technology Solutions Center, an HNTB center of excellence aimed at seeding digital infrastructure solutions throughout the company. Centers of excellence are an increasingly common unit at forward-thinking organizations.

Created by Siegel, a 27-year veteran of the company, the center provides consulting, advisory services, and implementation on many cutting-edge fronts across the company. Its members help other departments think through questions such as how to streamline an asset management solution or govern data better.

In a company of 5,000, the center employs 45 individuals who focus on specific "patterns," including infrastructure solutions, civil integrated solutions, asset management and resiliency, and geospatial and virtual engagement—the group Welch runs. As opposed to a Skunk Works group, they're a profit-making center, meaning their innovations must produce bottom-line value.

Because of that profit drive—and because so much of HNTB's work is grounded in location—GIS and geospatial intelligence are often the lens through which the team views challenges and solutions.

"Darin coined the term 'geospatial problem-solvers,' and that's the common thread we're looking for, whether it's a new GIS analyst that we just hired or a senior developer we recruited," Siegel says. "We're constantly looking for somebody who has that mindset of, 'OK, here's the problem. How do we solve it using the latest, greatest tools that are proven?'"

Capitalizing on the Wisdom of the Crowd for Tech Integration

Due in part to client demand, HNTB was already using digital twins on projects including airports and wastewater facilities. A coastal resiliency program for New York City had even

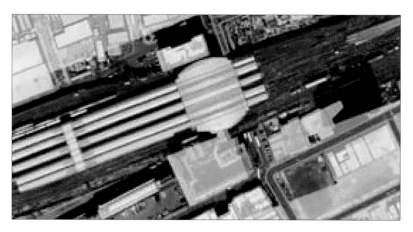

Two new tracks are being added to Jamaica Station.

employed 5D visualization. But Jamaica Station would be one of the company's most comprehensive applications of digital twin technology and 5D elements, incorporating both horizontal and vertical infrastructure.

To sketch out possible approaches, Siegel convened a lunch meeting at HNTB's Chicago office that ended up stretching across four hours. With boxes of deep-dish pizza on the conference table, the diverse team pulled up potential apps on a projector and filled dry-erase boards with scribbled ideas. They brainstormed what the best toolsets would be, how to enable them to work together, and how to make it all accessible via the web.

Welch's management philosophy—which reflects the interdisciplinary nature of GIS—is to bring together multiple points of view, and the meeting included specialists in several digital technologies. His team of 10 includes urban planners, transportation engineers, developers, senior analysts, and a geo-tech engineer.

"Diversity has always been important to me—bringing insights from lots of different backgrounds and interests to see how we can leverage these technologies to make our processes and deliverables more efficient," Welch said.

By the lunch's end, they knew GIS would be the hub of the wheel, fed by numerous software spokes responsible for replicating the surfaces, objects, and streets in the digital twin visualization. Although much of the design data was provided directly from the project team, a GIS-based "world building" tool gave the 5D immersive experience more real-world context by adding surrounding buildings and features.

An interdisciplinary team brought the vision to life. Senior GIS developer Ian Grasshoff became the visionary behind the web-based viewer; senior GIS architect Bill Cozzens helped design the realistic environment and city blocks around the rail's digital twin; and Scott Lecher, lead BrIM specialist from HNTB's civil integrated solutions team, helped align the visual elements into a smooth, scripted process.

To create the 5D view, the team built cost and calendar estimates from the ground up, taking advantage of the fact that each infrastructure element on the project had a cost and activity ID assigned to it. As the digital twin moves through time and a piece of a bridge pier or a track tie is set into place, that unit is added to the cumulative cost and tied to that section of timeline, synchronizing the three main elements of project management—scope, cost, and time.

With the digital model available to all stakeholders, planners can rally around a holistic view of the project, with communication grounded in a shared understanding of the plan.

"We've often referred to GIS as a natural integrator," Welch says. "We're using it as a way to visualize multiple components from nearly every discipline in a meaningful way. We're taking data, which these projects have vast amounts of, then turning it into information. But more important is turning that information into knowledge that supports decision-making."

A 5D Digital Twin Strategy Sets a New Standard

Siegel anticipates that projects such as this will raise the expectations for consultants such as HNTB. "I have calls every week about a client asking, 'I heard about this digital twin strategy. Should I be doing it?'" Siegel said. "We're seeing this concept really get a lot of interest."

He expects that in the future, elements such as 5D and digital twins will become part of other firms' offerings, even in

fields beyond architecture, engineering, and construction. For example, with executives paying more attention to workforce development gaps, visual, interactive twins could be key to transferring knowledge from retiring workers to new hires who need to become productive quickly.

Anytime visual, location-based projects need to be shared with stakeholders, a GIS-based digital twin can be ideal for translating data into clear, digestible 3D—or 5D—formats.

"It really is a big difference maker," Siegel said. "It's not a prettied-up rendering. This is real data coming from the design …. in an effective manner that's still visually pleasing and readable, but navigable as well."

New tracks are highlighted in green in the 5D digital twin of the Jamaica Station project.

Expediting Water Relief for the Navajo Nation

In April 2020, as the magnitude of the coronavirus pandemic and its impact were becoming readily apparent, Commander Ryan Clapp, a staff engineer with the Indian Health Service (IHS) in Washington, DC, flew to Albuquerque, New Mexico. Upon arrival, he bought eight pay-as-you-go cell phones from a retail store and loaded data collection apps on them. Within 48 hours, he had a team of Navajo Area IHS technicians spread out to map water access points on the trial nation using the mobile devices.

While he was in the air, IHS headquarters staff were developing a comprehensive field survey, talking to the Navajo Tribal Utility Authority, and doing all the background work.

"We were building things as we were going, and it was moving very fast," said Captain Ramsey Hawasly, assistant director of the Division of Sanitation Facilities Construction at IHS and lead GIS program coordinator.

This rapid response was requested by the Navajo Nation president because of the COVID-19 public health emergency. At the time, the Navajo tribe was experiencing the highest incidence of COVID-19 cases in the United States, and the long-standing lack of in-home water access was assumed to be a driver of these infections.

The heightened need for handwashing during the pandemic

Sandstone towers rise as high as 1,000 feet above the valley floor in Monument Valley, located within the Navajo Nation.

A total of 59 new water access points were created with CARES Act funding. (Image courtesy of the Indian Health Service)

Thev status of water access for each chapter on the Navajo Nation.

posed a challenge for the many homes without water. For many years, the rugged topography and remoteness of the Navajo Nation made piping water to homes challenging. Since 2003, IHS and a network of partners have reduced the number of Navajo homes without water access from 30 percent to 20 percent. New funding from the Coronavirus Aid, Relief, and Economic Security (CARES) Act provided the Navajo Area IHS with $5.2 million, targeted specifically to increasing water access on the Navajo Nation.

The IHS team used a GIS to map and share construction progress on new water access locations. Many are at or near chapter houses, which serve as county-level governments.

"We were able to collect data in all 110 chapters with just six surveyors, covering an area the size of West Virginia in just two weeks," said Captain David Harvey, deputy director of the Division of Sanitation Facilities Construction at IHS.

Immediate Short-Term Actions to Improve Water Access

Through the GIS data collection efforts, the CARES Act funds supported the installation of 59 new transitional water point (TWP) connections to existing public water systems; supplied 37,000 water storage containers; distributed 3.5 million water disinfection tablets; and subsidized the water for people living in homes with no piped water through February 2023.

The data that Commander Clapp and the team collected was critical in identifying locations for needed facilities. It set in motion additional mobile work to design each access point, calculate construction costs, and place orders for the right amount of pipe needed to make new connections.

A new water access point, now fully operational, was added near the Rock Springs Chapter house.

Building the Capacity to Act Together

Just four months before the start of the pandemic, IHS implemented technology that a group of GIS advocates had been excited to put into action for four years. This group took a change management class to learn how to prepare colleagues for significant process change.

"Luckily, we were already moving in the enterprise GIS direction before the pandemic hit," said Captain Shari Windt, engineering consultant for the Environmental Health Support Center at IHS. "We installed ArcGIS® Enterprise portal in 2019 and had been working to gain energy behind it."

The Navajo Water Project provided the opportunity to push digital workflows to the field using a suite of ArcGIS apps—ArcGIS® Survey123, ArcGIS® Collector, and ArcGIS® Dashboards—to equip field crews with the ability to collect data and provide updates on a central dashboard. The shared map led to new levels of collaboration.

"This project really helped the rest of the program see how beneficial this could be," Captain Windt said. "Rather than have information stuck in files or drawings, anybody with access can get to it."

The IHS made a digital leap forward in 2004 with the launch of the Sanitation Tracking and Reporting System (STARS).

This move consolidated databases that track infrastructure deficiencies in homes and communities; requests for water service to home sites; documents and details of operations and maintenance projects; and service requests. STARS serves as an inventory of tribal sanitation needs, and it put 400,000 homes on the map because the program focuses on serving homes in communities.

A new water access point, now fully operational, was added near the Rock Springs Chapter house. (Image courtesy of Indian Health Service)

"With the STARS system, we were able to access a lot more data and understand costs really quickly," Captain Windt said. "Now with GIS, we have the opportunity to combine all individual project drawings into composite drawings and make them readily available to IHS staff as well as our partners. We envision that the tribes will be able to leverage the GIS information gathered by the IHS to improve their operation and maintenance capacity, which support the water and wastewater facilities constructed with funding from the IHS and other federal agencies. It will allow the IHS to better understand each water and wastewater system as a whole, which will improve the technical support IHS can provide tribes."

The fast-paced construction lasted from mid-July through September 2020. It was guided and communicated through shared maps and plans at each step. An online interactive map marked progress for each chapter. In addition to the online map, a regularly updated map appeared in tribal newspapers because the newspaper is the primary source of information for many homes.

"We were changing colors on the map, based on whether or not there was an identified transitional water point for each chapter," said Captain Windt. The map changed when water points were slated for construction, design was being done, the construction was complete, and the water point was open. A final color was used when a chapter had all the interventions available.

The map guided the workers and the people they were serving to the new TWPs. Using GIS analysis, IHS calculated that the travel distance dropped from 52 mi. to 17 mi., saving people an average of 38 minutes behind the wheel for each trip.

Supplying Off-the-Grid Resources

Just as the water access work was wrapping up, the Navajo Nation Department of Water Resources (DWR) reached out to IHS under a separate request for help on bridging the gap for remote homes without access to a piped water connection.

DWR requested that IHS provide detailed design project drawings for a water cistern and on-site wastewater disposal facilities. Again, IHS stepped into action, deploying 15 Commissioned Corps engineers and environmental health officers of the US Public Health Service to undertake data collection.

"The cistern project is really beneficial because you can have bathrooms with showers and toilets, and sinks in kitchens," Commander Clapp said. "The homeowners still have to haul water, but it's a bridge to more sustainable services, such as a connection to a piped water system."

The data in the STARS system combined with analysis in GIS allowed IHS to identify 900 top candidates and provide a map of those homes.

Commander Clapp led three teams made up of five people each that spent a month in the field on assessments. For each home visited, the team gathered design data from the site to determine where the water storage tank and sewer facilities should go. Team members also assessed any potential problems, such as the location of large rocks or obstacles that might hinder construction or site access.

"It was a collaborative effort, and it was done in a quick and efficient manner," Captain Hawasly said. "The Navajo Engineering and Construction Authority could access the data collected to quickly create plans and drawings for approximately 70 homes."

This pandemic work spurred the creation of the Water Access Coordination Group, which includes four Navajo government entities, six federal government agencies, three universities, and two nonprofits. Bringing water to homes is something that everyone working on these projects feels passionate about.

"It shouldn't have taken this disaster to get us here, but now there's a whole new recognition of what tribal resilience means and how the federal government can work together," Captain Harvey said.

London: Maps and Location Technology Promote Mobility and Health

Twice in the last decade, Londoners have seen dramatic shifts in the way they interact with streets and transit. The first was in 2012 when the Olympic Games temporarily brought 600,000 new riders to London's buses and trains. The second was in 2020 when the COVID-19 pandemic emptied streets, buses, and rail cars. In both instances, planners at Transport for London (TfL) looked for near real-time understanding of traffic patterns, demand, and incidents using situational awareness from a GIS.

In response to the pandemic, TfL had to reconsider how mobility can enhance public safety. During the Olympics, it had to move people to many venues while maintaining mobility for residents and businesses. New technologies implemented in 2012 would prove valuable once again in 2020.

Olympics preparation included creating the Games Playbook, a comprehensive traffic management tool that served as a central source to visualize mobility. Reflecting on lessons from the Games, Michelle Dix, then director of planning

for TfL, told the BBC: "That's the biggest legacy in terms of behavioral change. We proved that through messaging and communications and telling people what's going on, telling them about alternatives so they can make informed choices, we proved that you can manage these big events."

GIS technology has given TfL a chance to understand how street space can be reimagined for London's residents—whether the challenge is to use the space efficiently for more people or use it safely for fewer.

The Olympics effort was deemed a success, with 90 percent of journeys completed on time despite a record number of riders. The London Tube alone had 4.5 million riders on one day of the Games compared with the typical 3 million, and 30 percent more than usual over the course of the event.

In addition to catalyzing the use of powerful traffic awareness technology, the Olympics kicked off a series of citywide initiatives to make healthy, sustainable travel options more accessible. Boris Johnson, then mayor of London, pledged to maintain key elements of the walking, cycling, and public transit infrastructure created to support the Games.

Now, public health is a top priority of Mayor Sadiq Khan's 2018 Transport Strategy, with the goal of 80 percent of trips around the city to be taken via public transit, on foot, or by bicycle by 2041.

TfL's GIS services have been instrumental in these efforts.

"After the Olympics, the need for geospatial understanding still remained very much at the forefront," said Jaymie Croucher, TfL's GIS lead for Network Management, Surface Transport.

London Bridge was lit with the Olympic rings during the 2012 Olympic Games.

TfL invested in its GIS in 2014, delivering GIS as a service and creating the Surface Playbook. "We're six years in to a 10-year plan, and we continue to grow to support internal and external stakeholders—all fed through the single source of truth for data," Croucher said.

Creating Space for Safe Streets during COVID-19

Since the emergence of COVID-19 in early 2020, TfL has faced a new challenge in support of social distancing and implementing London's goals for active, sustainable transportation.

Using maps of city streets created in the Surface Playbook, TfL used funding from the Government's Active Travel Fund to make more space for people to walk and cycle safely during the pandemic. The program's intention is "to support the members of the public to have more confidence to walk and cycle," Croucher said. "It's focused on supporting sustainable modes and increasing the ability to enact social distancing in transit, whilst limiting the impact of other modes."

The program encompasses several connected projects, including widening walkways, creating temporary bicycle lanes, and restricting car traffic near schools and in designated low-traffic neighborhoods. Projects have been implemented citywide on the 360 mi. of roads managed by TfL as well as locally through funding provided to London's boroughs.

As all the projects took shape, Surface Playbook was used as the focal point for giving clear situational awareness across the business as schemes were proposed, planned, and delivered. "There's a limited amount of space in London, and our team had to act quickly and use it wisely," Croucher said. "We needed to look at social distancing and find out where it was going to be the biggest issue."

The team's principal data manager and GIS specialists, Christophe Delatreche, Timothy Fegan, and Christina Kimbrough, along with the data scientists of Operational Analysis, created products identifying at-risk and high-demand areas of pavement to give planners a clear indication of priority areas. In particular, the maps highlighted five factors: areas with current high demand for cyclists and pedestrians, essential services such as grocery stores, population density, median household income (low income typically correlates to higher foot traffic), and high-traffic public transportation hubs. Together, the maps conveyed a clear picture of where streets were likely to be crowded—and where more space was needed most for safe and socially distant activity.

Traffic decreased signifigantly during the pandemic.

The resultant risk assessments could then be layered over data such as walkway widths to create priority scores. Top priority was given to streets and neighborhoods with both the need and capacity for wider walkways. "This enabled the business to have a clear understanding of precedence enabling clearer decision-making prior to accepting schemes." Croucher said.

Similar processes informed additional schemes, including added space for social distancing at heavily trafficked bus stops and establishing 24/7 bus lanes on priority routes.

TfL also coordinated with the GIS team within City Planning, led by Vicki Gilham, to expedite existing plans for cycle lanes created under London's broader transport strategy. Before the pandemic, Croucher said, "Where previously we had seen cycling programs paused pre-pandemic, we're now seeing those accelerated in areas where there's an increased need for cycling infrastructure."

Location Information at the Center of Traffic Management

The Network Management department at TfL delivers GIS as a service (GISaaS) to many departments and enterprise systems. The data and layers captured in TfL's GIS describe locations and modes of transportation, including up-to-date details about roadways, rails, paths, and all the physical assets TfL maintains. Sharing this capacity as a service means other systems can build on authoritative data to visualize, query, and analyze it for specific purposes.

TfL undertook a major digital transformation in 2016, with its Surface Intelligent Transport System (SITS), an umbrella project that modernized traffic signals, incident management, and the coordination of road improvement.

TfL's GIS as a service directly feeds its adaptive traffic signal system. Data from sensors in the road network feed the split cycle offset optimization technique (SCOOT) model that analyzes volume second by second. Buses and their location in the queue are also modeled and monitored. Then, the model feeds junction controllers in real time to adjust traffic signal timing to take advantage of each road's volume versus capacity and to coordinate flow with neighboring roadways.

TfL's operational knowledge, captured in GIS, helps improve traffic flow and the environment for walking and split-cycle. One of the traffic signal innovations tested during COVID-19 involved "trialing the Green Man Authority," where pedestrians see a continuous green man until a vehicle approaches.

GIS and the SCOOT system also feed an advanced big data analytics tool to understand dynamic changes in road traffic. The Real Time Origin Destination Analysis Tool (RODAT) analyzes feeds from video cameras at key locations and along major routes in central London. The system monitors more than 20,000 origin and destination pairs every 15 minutes to calculate actual journey times and traffic flows to keep London moving.

GIS is also at the center of TfL's LondonWorks system for coordination of road improvements, with a registry of all roadworks and street-related events, both planned and current, in the Greater London area. LondonWorks maps all incidents and uses spatial analysis to assess road networks and then coordinate various roadworks to minimize congestion.

All inputs are combined in TfL's Traffic Information Management System (TIMS) to monitor and manage traffic using a GIS database of live and planned traffic disruptions in London, including congestion, traffic incidents, repair work, and events. TIMS allows media agencies and other stakeholders to view disruptions in real time (updated every five minutes) and see information about planned activity that's likely to impact traffic—providing a shared situational awareness.

Mapping the Future of London Transit

Although the efforts align with London's long-term goals, many schemes introduced during the pandemic are currently considered temporary. TfL is working to report on the success of each scheme to gauge whether the changes will make sense for the city after the pandemic—much like the process for maintaining Olympics infrastructure following the Games.

"Before we can decide whether a feature becomes permanent or not, we need to understand what the impact is," Croucher said. "If you close a street, for example, that is going to cause traffic to develop elsewhere. The schemes need to be well monitored to ensure that they're effective in the long run for all modes of transport."

On-site surveys have been a key component of the monitoring process. Surveyors in the field across the Greater London area capture data about use, safety, and needed improvements. Survey feedback populates a live, online GIS dashboard, providing real-time visibility for city management and informing next steps.

To that end, such shareable reporting tools have supported TfL during the pandemic. With many schemes also being carried out by boroughs, the team created a GIS database to centralize program information.

London's congestion zones charge a fee to discourage travel by car in the center of the city.

"Compiling it all within a single portal, we've allowed everyone to have transparency over the safety of each scheme and compliance of users," Croucher said. The portal delivers situational awareness for TfL and city leadership to see how each scheme performs and interacts.

Changing a city's transportation habits ultimately requires a strategic rebalancing of the way residents use city streets. Recreation, public transit, and the transportation of people and goods—each function claims a space on city streets.

Giving more space to one will necessarily take space from another.

Maps provide a powerful visual understanding of the space available and a strong platform to plan, prioritize, and improve.

"The benefits of the way we collect and disseminate this information are reaped well beyond TfL to provide clarity to both external partners and, ultimately, the public," Croucher said. "Understanding the spatial relationships that elements have provides a clear picture for decision-makers that you won't necessarily see by looking through more traditional means such as a database or a spreadsheet."

Hampton Roads Gains a New Operational Awareness to Improve Wastewater Operations

In southeastern Virginia, Hampton Roads Sanitation District (HRSD) handles wastewater treatment for 1.7 million people in 20 cities and counties. Sea level rise, unusually high tides, and extreme storms prompted a $1.2 billion program, the Sustainable Water Initiative for Tomorrow (SWIFT), that involves replenishing the Potomac aquifer with up to 100 million gallons of SWIFT Water™ (water treated to meet drinking water standards and matched to the existing groundwater chemistry in the aquifer) per day. Replenishing the aquifer may slow or reduce the impact of sea level rise by slowing land settling, or subsidence.

"Geologically, sea level rise in our area is not just about water warming and ice caps melting," said Anas Malkawi, chief of asset management at HRSD. "Those are factors, but it's also due to the overdraw of groundwater that causes the land to sink."

The SWIFT solution will recharge the overdrawn Potomac aquifer with water treated to meet drinking water standards and match existing groundwater chemistry. The work will renew a resource that has been greatly depleted over the past 100 years, arresting subsidence, adding resiliency to handle stormwater events, and combating the growing issue of saltwater contamination.

SWIFT and other innovative approaches are indicative of HRSD's leadership. Well established in the area, HRSD has provided regional wastewater processing capacity since 1940. It operates 17 wastewater treatment plants collecting from pipes and pump stations run by local governments near the mouth of the Chesapeake Bay—and their work scope is growing because of environmental and economic pressures.

"Smaller communities can't afford to meet increasingly stringent water quality requirements because they don't have the population and can't raise their rates high enough," Malkawi said. "HRSD has a good bond rating to borrow

The digital twin of HRSD's SWIFT facility includes attribute details that can be explored and queried. The exterior *(left)* includes details on pipes underground whereas the interior *(right)* includes details about specific equipment.

funds, distribute the load, and upgrade the infrastructure across the whole region."

Introducing a Digital Twin

As a cutting-edge climate action project, SWIFT provided the impetus for HRSD to employ new technologies, working with engineering firm Hazen and Sawyer.

First, they combined building information modeling (BIM) designs with GIS maps, workflows, and analysis. Together, BIM and GIS data helped the district establish a feature-rich model, known as a digital twin, of the SWIFT Research Center that provides the proof of concept for the program. Over the next 10 years, HRSD will build at least four more facilities to resupply the aquifer with up to 100 million gallons of water per day.

Now, planners can explore the design models via virtual and augmented reality, wearing goggles that immerse them in the infrastructure. Operational staff and construction crews can see where a new pump or asset will be located, for instance, and ensure that it gets placed in the correct location. Facility owners can assess the maintainability and operation of assets before construction. And the models provide an ideal training platform for new staff.

"Seeing the full context of the plant in the design phase really helped us see the value it can bring to operations and asset management," Malkawi said. "The ability to navigate through the plant with access to operational and maintenance data in one platform provides great value for our workforce."

The digital twin is affording a higher level of operational intelligence with 3D data for newly constructed facilities and plans to capture the same for older facilities. Although they are currently gathering and visualizing real-time sensor data, the team will eventually move to automating systems based on sensor input.

"We've achieved the standard value from our digital twin of more situational awareness," Malkawi said. "We would like to simulate our system, looking at the behavior of the infrastructure and knowing, if I open a valve and change a process, what are the effects to the downstream process?"

Sensing and Examining Changes

To help ensure water quality, HRSD crews test private and commercial irrigation wells near where treated water recharges the aquifer. They use sensors to measure shifts in water quality, syncing sensor data on a GIS map where it can easily be monitored.

The district takes a similar, sensor-driven approach to monitoring pressure and pump performance for its 651 mi. of pipes and 131 pump stations. Staff keep an eye on their GIS interface to examine assets at every location. The data feeds an operational dashboard with a map, sensor readings, and graphs to track variables such as salinity.

Vertical and horizontal assets mix within the treatment plant.

"We have conductivity meters on the different sections of our gravity system," said Jules Robichaud, GIS manager at HRSD. "The historical data is one click away. In areas at high risk for sea level rise, we're trying to get a sense of where high salinity may be an issue."

When storms hit, the HRSD team works to reduce wastewater overflows and monitor infrastructure. With increasing storm intensity and the compounding factors of subsidence and sea level rise, there's greater urgency to protect wastewater drainage systems from spills. For this effort, the district will again lean on its GIS and digital twin technology.

"We have to make over a billion dollars of infrastructure investments," Malkawi said. "We have a consent decree with the US Department of Justice and the [Environmental Protection Agency] EPA that we have to meet certain targets to reduce or eliminate sanitary sewer overflows to a certain rain event level. Incorporating risk into the digital twin is going to be a huge value."

Modern GIS for Operational Intelligence

For more than 12 years, the district has relied on GIS as a critical tool for day-to-day operations. In addition to smart maps and dashboards, GIS provides analytics to help HRSD meet growing infrastructure needs.

"Our operators have GIS at their fingertips to navigate to linear assets like pipes, find information, collect information, and report it all out via a map-based tool," Malkawi said. "If we need to know where the asset is in proximity to businesses, residences, local water bodies, and traffic, we use GIS. If there's a failure on some of our infrastructure, we use GIS to understand the impact on the public, on worker safety, and [on] the environment."

Many of these capabilities have been aided by the modernization of GIS to deliver and consume data as services.

"We're bringing in millions of records of data from multiple jurisdictions," Robichaud said. "We automate scrubbing that data, combining it into regional layers, and making it available to regional partners."

The district's water quality team uses GIS to track pollutants and pathogens to their source—an endeavor made easier by the integrated data environment.

"Having access to regional data for the jurisdictions we serve, the stormwater and sewer layers, and even parcel data to locate businesses, industries, and communities has been huge," Robichaud said. "We tie lab results back to where samples were collected, and then share that with the Virginia Department of Health and the Department of Environmental Quality."

HRSD started its use of GIS for operations, interfacing with its computerized maintenance management system (CMMS), and for managing customer information. "We have over 70,000 assets in our organization that all require routine maintenance," Malkawi said. "Whenever we do a repair, we record what was done, what caused the failure, and other details." GIS is used to learn the age, condition, and performance of the asset as well as to monitor risk, hydraulic capacity, and processes.

"We've used 20 years of GIS data to make maintenance decisions," Malkawi said. "Looking at our pipeline failures, for example, we can do a hotspot analysis using GIS tools to see the primary areas of failure, what's causing failures, and figure out what we need to do with the rest of the pipes that have similar characteristics."

The first facility for HRSD's Sustainable Water Initiative for Tomorrow (SWIFT) project serves as a demonstration site.

Along the Mekong River, Development Creates Sustainability Concerns

Rapid growth in the number of dams along the Mekong River is transforming Southeast Asia's energy, food, transportation, security, and ecological networks. Eleven hydropower dams now span the river's mainstream before it leaves China, with hundreds more dams planned or under construction in the other countries that contain parts of this vital watershed. The development has been rapid, and the consequences to people and the environment are still largely unknown.

Scientists at the Stimson Center, a think tank focused on enhancing international peace, have been monitoring the hydroelectric power projects in terms of impact on regional stability and the food-water-security nexus.

"Our core message is that nonhydropower renewable energy, like solar and wind, can replace hydropower with far less disruption," said Brian Eyler, director for Southeast Asia at the Stimson Center. "Our static map of Mekong mainstream dams is very popular for use in the media and for research presentations, but we realized that a static map didn't show the full picture."

With support from the US Agency for International Development (USAID) and the Asia Foundation through the five-year Mekong Safeguards activity, Eyler and his team set out to create the Mekong Infrastructure Tracker, a database of visual presentations, spatial analyses, and shared expertise. The dashboard was created using a GIS, with help from Esri partner Blue Raster.

"In addition to tracking dams, we had the idea to track solar projects after seeing a big uptick in the region in 2018," Eyler said. "Our partners at USAID asked us if we'd track all infrastructure in Cambodia, Laos, Myanmar, Thailand, and Vietnam. It didn't take much consideration to jump at this opportunity because infrastructure is a hot issue in the Mekong region."

Xayaburi Dam on the Mekong River in Laos.

The first bend of the Mekong River as it emerges from the Tibetan Plateau in China.

The tracker catalogs projects for power generation and industrial development as well as road, rail, and waterway transportation. It also includes tools to analyze and quantify the impacts of infrastructure projects.

Construction on the existing dams has already displaced thousands of people, causing immediate impact for some. Other river people who subsist on fishing and riverbank agriculture are feeling the effects of the ecological damage to the ecosystems. In the village of Baan Huay Luek in northern Thailand, for example, villagers speak about how the dams have switched the rules of nature, with dry seasons no longer dry and wet seasons no longer wet.

Mapping Mekong Infrastructure Development

The Mekong River starts on the Tibetan Plateau in China and flows through Myanmar, Laos, Thailand, Cambodia, and Vietnam before emptying into the South China Sea. The river has been deemed Southeast Asia's most important waterway as it is central to the lives and livelihoods of millions of people and serves as a food source. The Mekong watershed is known as the "rice bowl" of Asia, and 20 percent of the world's freshwater fish catch comes from its waters.

To date, governments and activists interested in Mekong watershed development have had to contend with the lack of information about, or awareness of, existing or planned projects.

"This information is usually something that private-sector organizations collect and then distribute only to their clients," said Regan Kwan, research associate at the Stimson Center and manager of the Mekong Infrastructure Tracker project. "Through GIS, we make the data available so that anyone can visualize it, analyze it, and contribute to it."

The Stimson Center works with a wide range of stakeholders on this and other projects, with an eye on community engagement and capacity building. The Mekong Infrastructure Tracker has drawn interest and participation from national development banks, government agencies, ministries, international nongovernmental organizations, local grassroots organizations, the private sector, academic researchers, and individuals.

"Our Stimson team is relatively small, and we're collecting thousands of data points," Kwan said. "We're bound to make some mistakes, skip things, or not see everything others are seeing. It's great to have others use our data, share their viewpoints, and fill in gaps."

In addition to serving as a singular accurate data source, the Mekong Infrastructure Tracker helps guide foreign policy and assess investment opportunities.

"Having all this information transparently and freely available gives those looking to achieve a more sustainable future a resource to ask questions anytime they want," Eyler said. "No one has provided this depth of information before."

Learning about Energy and Environment in the Mekong

The Stimson Center avoids advocacy, instead allowing partners to draw their own conclusions from the data. One such partner—EarthRights International, based at the Mitharsuu Center in Chiang Mai, Thailand—trains young activists on development issues, environmental impacts, and human rights law and policy. During a seven-month program, called the EarthRights School, students take classes and learn to conduct research.

"We've been encouraging our students to use the Mekong Infrastructure Tracker to map out various projects in their home countries in order to support their advocacy," said William "BJ" Schulte, Mekong policy and legal adviser at EarthRights International. "We value the Tracker as a strong tool that can support networks of activists and community leaders to address the regional energy strategy in the Mekong region as we see a growing reliance on fossil fuels, especially coal."

One of the reasons fossil fuels are gaining interest in Cambodia, for example, is that existing dams haven't produced as much energy as predicted, largely because of an ongoing drought. Coal causes grave concern because it releases the most greenhouse gas of any fuel, and the hazardous waste by-product coal ash is hard to safely store.

"One of our alumni alerted us to a coal-fired power plant that's going to be sited within [Preah Monivong] Bokor National Park," Schulte said. "We're working with local activists to understand the local impacts."

EarthRights International has participated in adding data from communities to the Mekong Infrastructure Tracker to raise awareness about such projects.

"The tracker gives us the ability to draw out where all the projects are and see how they impact each other," said Naing Htoo, Mekong program director at EarthRights International. "We use it to see the regional perspective, including the financiers behind the projects, which includes a lot of cross-boundary investment."

Gaining Geopolitical Awareness

Many government leaders are seeing the Mekong Infrastructure Tracker data and getting a greater regional perspective.

"At first, we had some anxiety about presenting our data to governments in the region, because satellite imagery and other forms of remote sensing data are often seen with suspicion," Eyler said. "We found that data which shows the speed of change in the region and the ease of visualizing and analyzing the data appeases concerns. It was the first time they could see what's happening in other countries."

China's ambitious Belt and Road Initiative plays a major role in much of this development. The Mekong Infrastructure Tracker promotes awareness of this and other development programs.

"The data can be used to assess the geopolitics, seeing areas where China, or other external investors from Japan, Korea, or the United States, is more heavily invested," Eyler said. "You can see how many projects are planned and how many are moving

The Mekong Delta flows through Myanmar, Laos, Thailand, Cambodia, and Vietnam before emptying into the South China Sea.

forward. Knowing the gaps is really useful information."

With more infrastructure projects being planned, and with the rising risks from climate change, EarthRights International workers are also seeing more negative impact on Indigenous peoples.

"Due to their locations, many of the large infrastructure projects disproportionately impact ethnic minorities and Indigenous peoples," Schulte said. "When they try to raise their voices, they are increasingly attacked or imprisoned, or even disappear. A lot of the fossil fuel projects are backed by powerful people with significant financial interests."

Operationalizing Data Gathering and Shared Insights

The transparency of the Mekong Infrastructure Tracker— along with the elements of citizen science that allow anyone to contribute—is encouraging widespread participation from a diverse array of people. The Stimson Center takes a data-accurate and research-agnostic approach to the platform to engage participants and encourage dialogue. Administrators are driven to gather complete data and present the facts.

"If anyone using our geodatabase realizes it differs from what they know about an infrastructure project, and they have sources to back it up, we provide an ArcGIS Survey123 form to fill in missing information," Kwan said. "We review it, and if it checks out, we make the update."

The Mekong Infrastructure Tracker has a Facebook group that participants use to drop links to news reports about infrastructure projects. If the news signals a new project, it's added to the database. Any information about changing timelines or project details is noted. In addition, hackathons are held to fill in data gaps and correct inaccuracies.

The dashboard focuses on five main datasets for energy, transportation, and water infrastructure projects and has data about the surrounding environment as well as people, including ethnic groups. Data on energy projects—including biomass, coal, gas, geothermal, hydro, mixed fossil fuel,

nuclear, oil, solar, waste, and wind—is curated along with transportation details on upgrades to canals, urban and high-speed railways, and national roads. The site also covers industrial development at airports, railway stations, and sea or inland ports, as well as special economic zones that provide investment incentives.

"We work to ensure the data is accurate and complete," Eyler said. "The vision is to continue to provide a mix of technical and simple-to-use tools to analyze the data and improve decision-making in the region."

The Mekong Infrastructure Tracker database provides robust environmental and social impact indicator layers that allow anyone to see and study the risks and benefits of various development scenarios.

"A long-term goal is to create a scenario planning application which gives users a chance to generate their own investment scenarios and assess costs and benefits across a variety of indicators," Eyler said. "Users could compare scenarios and even take their scenarios to other policy makers and planners to talk through multiple development pathways and choose the most optimal for their own locality or for the region at large."

The Mekong Infrastructure Tracker database builds on existing data to present a comprehensive source of information on energy, transportation, and water infrastructure in the Mekong countries.

Public Safety and Humanitarian Response

The latest iteration of GIS for public safety involves an integration of tools and real-time inputs to maintain situational awareness and enhance the flow of information for responders. A modern GIS combines data collection, analysis, and sharing to achieve operational intelligence.

True operational intelligence helps managers see and communicate about incidents and changing conditions. With this system, the public safety community gains a deeper understanding about where to focus efforts to prevent, protect against, and mitigate the effects of complex threats and hazards. It allows agencies to understand, plan, and act together.

An optimal system includes up-to-date details about community residents to quickly connect responders to 911 calls, with routing to the right address. In Illinois, the state police have embarked on an integrated statewide Next Generation 911 system that connects all 911 call centers, using GIS to improve response times. Modernization includes the ability to receive photos and videos from callers with smartphones to improve responder preparedness.

Anticipatory action is one of the latest buzzwords in the response sector. The thinking is simple: if you can plan for the disaster before it hits, you can save more lives and save money. GIS helps with that planning.

Rising global temperatures have intensified both rainfall and drought, leading to too wet and too dry conditions that heighten the risk and impacts of disasters. These extremes were on display last summer with devastating floods and fires in Europe and the United States that put vulnerable people and places in the spotlight.

When severe weather events such as high winds and floods impact the same location over and over, they leave residents more vulnerable and consistently in recovery mode. But a better understanding of places can lead to mitigation or adaptation strategies. Emergency management and disaster response planners emphasize the importance of understanding an area's systemic risk to weather and environmental conditions to break the familiar cycle of respond, recover, and repeat.

As the climate changes, more extreme and frequent disasters are hitting communities that haven't previously had to deal with catastrophic emergencies. For instance, Tropical Storm Irene caused inland flooding in Vermont, destroying more than 100 iconic century-old covered bridges—structures not built to withstand a deluge of that force. Similarly, the floods in Germany in July 2021 were the result of record rainfall that led to flooding that swept away homes, forced evacuations, led to widespread power outages, and devastated agriculture and centuries-old infrastructure.

In places frequently struck by hurricanes or tornadoes, preparedness is a way of life. However, extreme storms brought on by climate change are impacting communities unaccustomed to disasters. And emergency managers are increasingly tasked with the responsibility to coordinate the response to natural, technological, and human-made hazards as humanitarian response grows in the face of growing disasters.

With the focus on anticipatory action, it is regional emergency planners who will make the greatest impact with their community's share of increased funding for preparedness. They will need GIS to understand local risks, see where mitigation projects will have the greatest impact, and figure out how to equitably address community needs.

Location Intelligence Empowers Aid Organizations to Act as One

By Ryan Lanclos, Esri director of Public Safety Solutions

As countries around the world continue to feel the effects of COVID-19, a new level of communication through shared context is helping people across organizations and governments work together to curtail the pandemic.

One source of truth, the Johns Hopkins COVID-19 Dashboard, uses Web GIS technology to show the global status of the spread of the virus. This location intelligence, based on Web GIS, connects and informs people who are taking on COVID-19 challenges.

Government officials use GIS for a deeper understanding, analyzing the vulnerability of people and monitoring the effectiveness of policies designed to slow the spread. Business leaders rely on GIS to analyze facilities, supply chains, and customer activity to know when and where to reduce or ramp up activity to weather the pandemic. Staff at global nonprofits perform spatial analysis with GIS to guide the much-needed delivery of aid.

These efforts are all supported by Web GIS, which delivers a combination of location-aware apps, dashboards that display critical dimensions, spatial analytics to visualize what's

When the US Army Corps of Engineers completed the setup work at the Javits Center in New York City, the Army and Air Force Exchange Service helped staff the center.

happening across space and time, and maps to guide actions. This geospatial toolset strings together workflows visible to all. Web GIS has been put through its paces to tackle many compelling missions because this era calls for unprecedented collaboration. The technology pattern has matured at the right moment with enhanced capabilities—including app-based workflows, artificial intelligence (AI), real-time analysis of Internet of Things (IoT), and deep learning—that make it easy to link data and combine efforts.

Answering Rapid Requirements

The US Army Corps of Engineers deployed its first shared enterprise application across all 52 districts nationwide to connect workflows and share progress against the goal of adding hospital beds. The corps usually responds to much more localized disasters. The scale of the COVID-19 crisis forced a new level of focus, with each district using the same tool to assess facilities that could house hospital beds.

A shared national mission, on a tight timeline, brought together two key web-based geospatial capabilities. To prioritize relief efforts, corps leadership needed real-time information and spatial analytics—forecasts and projections of disease spread and hot spots. Crews fanned out using field mapping to audit and assess potential locations for additional beds, such as hospitals, college dormitories, and public arenas.

Crews gathered details, powered by location-aware apps, to amass expertise from multiple specialists for each site audit. The app gathered input from structural and civil engineers, environmental specialists, and other technicians. This information was aggregated and analyzed alongside site specifics such as proximity to public transportation and other hospitals, as well as distance from distribution sites. All the data was fed into the GIS for spatial analysis to give each site a score. Then, as sites were selected, the same GIS tools could track construction progress. To forecast future needs and guide planning, analysts applied ArcGIS® Insights℠, a web-based analytics engine that aggregates inputs from multiple models to guide priorities.

In short order, the alternate-care facility operation resulted in more than 600 assessments, 38 construction projects, and more than 15,000 beds. The US Army Corps of Engineers had recently upgraded to apply ArcGIS technology across its enterprise when the pandemic hit. This advanced level of geospatial capability provided a proving ground and a path forward for further digital transformation.

Connecting Food and Peace

The World Food Programme (WFP) won the Nobel Peace Prize in 2020 for its efforts to combat hunger, its contribution to bettering conditions for peace in conflict-affected areas, and for acting as a driving force in preventing the use of hunger as a weapon of war and conflict.

As COVID-19 exacerbated the need for supplies worldwide, WFP extended its use of ArcGIS to bring aid to extremely dangerous and hard-to-access communities.

For example, the WFP's Afghanistan operations posed significant challenges. The country's protracted conflict, rugged terrain, rough road network, and snow and floods made it difficult for aid workers to deliver food to remote areas—often places with the greatest needs.

Easy-to-use location-aware apps have provided a new level

of awareness for people working on these difficult operations. Warehouse workers aggregate input from drivers about road conditions and enter it in a system that displays security threat, weather, and program details. The shared understanding has helped WFP staff forge greater alliances with locals in areas they once struggled to reach.

WFP leaders make heavy use of GIS technology to plan operations, map vulnerabilities, and meet the demands of the populations they serve. The tilt toward online tools enhances existing data collection, analysis, and mapmaking efforts by adding shared context. This understanding gives all stakeholders an awareness of the challenging dimensions of WFP's mission as the technology supports collaboration and iteration of solutions to achieve results.

Testing Connections

The volunteer organization GISCorps uses Web GIS to harness the skills of the professional GIS community to aid those in need. Operations include capacity building in developing countries. Recently, volunteers have been empowering a distributed network of professionals to apply local aid to communities in crisis. The work involves developing maps and applications for nonprofit partners as

The World Food Programme faces heightened food distribution challenges in Afghanistan during winter months.

A testing swab is offered to test for COVID-19 at a drive-through checkpoint.

A signature of Web GIS is its ability to connect disparate groups, enabling them to act together from remote locations. This flexibility supported staff across businesses, nonprofits, and government organizations who had to quickly pivot to working from home without losing productivity. The same connection drives the GISCorps testing sites map, bringing together local GIS users who gather data from authorities or drive by a location to make certain it's in operation and then add this knowledge to the map.

well as creating maps of crowdsourced photos for emergency managers during and after natural disasters.

The network of volunteers rallied early in the pandemic to collect and map details on US COVID-19 testing sites. The network created an online form that providers and government agencies could use to add their own testing sites. The resulting interactive map filled an information gap at a crucial time.

For the GISCorps testing sites dashboard, a web services approach connected disparate data with linkages that can appear on multiple maps. The famous Johns Hopkins dashboard takes the same approach—showing people around the world what is happening and where. This new level of data fluidity is important in modern dashboards that compile authoritative resources in one place. More than ever, people are relying on and seeing the value of shared understanding in a crisis.

The global pandemic is accelerating the evolution of location intelligence as GIS technology enables more collaborative, insightful, timely, and impactful response. The delivery of testing, protective equipment, food, disaster response, and vaccine distribution are all benefiting from higher levels of connectivity.

In pace with the changing needs of people and organizations around the world, Web GIS continues to evolve, adding capabilities at ever greater speed and scale. The connection capability clearly shined in this latest crisis. The latest GIS advancements include the combination of drone data capture, realistic modeling, and immersive augmented reality to provide needed information. GIS users can expect a more immersive contextual experience that improves humanitarian communication and collaboration for solving complex problems.

American Red Cross and Partners Work from One Map to Reach People in Crisis

As millions of people were out of work and many more were going hungry, emergency food assistance organizations saw the need for aid increase by as much as 500 percent during the last year. The accelerated demand, inflated by the pandemic and recent disasters, put an added strain on nonprofit groups. And now, an American Red Cross online data hub enables a new level of coordination.

"This revolutionary tool for disaster use is changing the landscape of coordination and collaboration among response and recovery organizations," said April Wood, senior director of External Program Services, Disaster Operations, at the American Red Cross National Headquarters in Washington,

DC. "Red Cross is working with partners such as Feeding America, World Central Kitchen, Southern Baptist Disaster Relief, Salvation Army, and Operation BBQ, to name a few of the organizations actively involved."

"They're used to having to swing for the fences sometimes," said Patrick Colley, director of member engagement at Feeding America. "But sustaining that for months and months on end has been a real challenge."

Feeding America is a nationwide network of 200 food banks that provide food assistance to more than 40 million people through partner food pantries, soup kitchens, shelters, and other community-based agencies. The number of people helped has grown as the COVID-19 crisis persists.

"We know that many of our members throughout the US have been adding one to two facilities per week, either for temporarily storing the food they need to meet the increased demand or spread people out to social distance enough to allow production to continue," Colley said.

Arriving at a Shared Awareness

To support Feeding America and other nonprofit groups, corporations, and federal and state agencies, the American Red Cross built a new Disaster Partner Hub, thanks to the support of Walmart and the Walmart Foundation. It brings together more than 70 organizations to share information related to four key areas: feeding operations, call center data, damage assessment, and shelter operations. A pilot project ran from April through June 2020, focused on pandemic food insecurity. Based on that success, feeding partners have incorporated use of the hub into their daily operational workflows.

"Feeding America has been an early adopter that has been in lockstep with us since the beginning of the vision," said Wood. "We were able to pull together data from partners in an automated fashion to visualize daily feeding activity in one dashboard for the first time."

Capabilities of the American Red Cross Disaster Partner Hub come from the organization's GIS, a technology that delivers online maps and apps for preparedness, response, recovery, and planning.

In building its enterprise GIS solution, known as the Red Cross Visual Interactive Event Wizard (RC View), the American

Feeding America has transitioned to drive-through food pantries during the pandemic to practice safe social distancing.

Red Cross fostered a digital transformation. RC View is used regularly by the 70,000 members of the Red Cross team, which represents a large portion of its 20,000-person workforce and 372,000 volunteers.

The American Red Cross Disaster Partner Hub provides access to GIS capabilities to aid partners. This data sharing extends beyond pandemic food insecurity response and into collaborative efforts to help recent hurricane victims.

"It's given us a whole different way to communicate," Colley said. "We're communicating now with data and seeing the hurricane areas a lot clearer. It serves as an immediate check to see where we're responding, quickly with a glance, and it lets us share status updates with the entire group and decide where to best direct resources through good communication."

Food Distribution Apps and Analytics

The American Red Cross has created several apps to address food distribution operations during disasters, carried out with other nonprofit and corporate partners. The apps help everyone quickly see who's doing what and where.

The Feeding Resource Reporting app details where partner agencies are responding and what their capacity is at each site. This information populates the Daily Feeding Activity app that shows efforts across the US, which can be sorted and viewed by disaster. An interactive map allows users to zoom in and see current activities and soon will include the near real-time tracking of food distribution vehicles.

All this information feeds a more analytical product called the Partner Feeding dashboard, which shows meals produced, meals served, and food boxes delivered across the country. There are also details that compare delivery versus capacity, which helps identify gaps and areas of need.

"This visibility allows us to navigate across department and organization lines—the subconsciously drawn silos of information—to be helpful," Colley said. "We can see what each other is trying to do and can fill in any gaps with resources."

Picking Distribution Points

Although COVID-19 case numbers can escalate rapidly, the

Serving up and delivering food to disaster survivors was easier before COVID-19, when the proximity of people wasn't a key planning consideration.

Partner Feeding Summary Dashboard shows meals the Red Cross served after Hurricane Sally in Alabama in 2020.

assistance officer at the Florida Division of Emergency Management. "The central tool will make our response better. In addition to understanding where current efforts are under way, analysts at the American Red Cross can now look at historical disaster response data to score regions for their risk propensity."

The analytical information is helping the American Red Cross and partner groups prepare and respond to an unprecedented combination of sequential disasters, along with a global health crisis.

need for assistance resulting from the disruptions caused by the disease often lags. Although there is no shortage of need, it can be difficult for aid organizations to determine where and when to focus their attention.

"The day they test positive isn't the day they need food," Colley explained. "Seeing indicators on a map helps us be mindful of not just where the hot spots are right now, but which of these hot spots are going to need continued support afterwards."

COVID-19 adds to the complexity of disaster response because the crisis has coincided with some of the largest wildfires and hurricane events the nation has seen, from the western states to Florida.

"The wildfires that have been plaguing the West Coast are not the same as hurricanes because people will weather a storm, but you can't weather a fire," Colley said. "Instead of a pinpointed place to bring aid, there is an immediate population movement, and we need to take the food to the people, which can sometimes be several counties or states away."

"The Partner Hub is a game changer as far as data collection and visualization," said Doug Roberts, state individual

The fact that the hub is delivered as a software-as-a-service implementation and accessible anytime, anywhere has been crucial for many partners working from home and using mobile devices. In addition, American Red Cross has leveraged the ArcGIS StoryMaps app to produce more than a dozen event-specific Partner Situational Awareness Briefs that have been viewed more than 22,000 times over the course of eight months.

"The ability to connect virtually through the hub has become an important space for us and our partners to work in during the pandemic," said Nigel Holderby, director of Disaster Public Affairs at American Red Cross.

The GIS tools have fostered more direct connections between the individuals and organizations working to help people in times of crisis.

"It's really amazing," Colley said, "to see how all of your coworkers, community, donors, other national partners, when they can clearly see the exact same goal and the same need and the path toward it, how amazingly collaborative and supportive we can be."

Fire Management Analyst Tackles Wildfire in the Field and in the Office

Casey Teske sees her mission statement as "Understand fire."

It's that simple, and that complicated.

For Teske, a fire management analyst for the US Fish and Wildlife Service Branch of Fire Management who has extensive experience in both fighting wildfires on the ground and understanding the science of its actions, that breadth of understanding is vital. It's the reason she's often the bridge between those who fight a tactical wildfire battle in the field and those who analyze the behavior of the blaze in the office.

As the 2021 wildfire season began, early fires underscored the vast variety of interlocking variables that lead to fires. Predictive analytics suggested a very busy fire season—and they weren't wrong. By late June, there were 29,362 wildfires that ravaged 1,164,555 acres. That's up from 22,817 fires and 1,044,248 acres at the same time in 2020, according to the National Interagency Fire Center.

Teske stays on alert with concerns about affected ecosystems, nearby residential communities, and the firefighters who risk their lives. She's ready to go anywhere at

a moment's notice to offer her insight, often gained using GIS technology.

Fire is both a potential destroyer of wildlands and a necessary tool for conserving precious ecosystems. The paradox is part of what has intrigued Teske from childhood, as she learned about the interactions of natural processes from her father, a biologist.

"Fire is really important," Teske said. "Species depend on it for survival. It can also be used as a tool: if you need to knock back the fuel with small controlled fires so that you don't get big explosive wildfires threatening communities, prescribed fire gives you a way to do that."

A Journey from Firefighter to Researcher and Back

It all makes sense now, but Teske's journey to that rare spot, as a link between the practical and theoretical sides of the equation, was a path she look by following her interests.

Near real time mapping of incidents including heat detection, lightning strikes, and infrared-based fire perimeters *(left screen)* can help analysts calibrate fire models to give more accurate fire behavior and spread outputs *(right screen)*.

"When I got out of high school, I decided to go to college, and I could fight fire during the summer to pay for college because they make good money, and I wanted to stay in shape," Teske said. "So, I signed up to be on a fire crew, and that first year was pretty pivotal for me because I learned a lot about a lot of different aspects of fire, and I got pretty interested in the fire behavior side of things."

She kept returning for several summers to the fire crew as her fascination with firefighting grew. She even decided to pursue a master's degree in fire ecology. It was then that she first learned about GIS and the power of location intelligence.

Meanwhile, she continued fighting fires in the summer to earn money to pay for her studies, and she kept thinking about how to link firefighters in the field with the insights that GIS could deliver. The more Teske knew about both sides (she eventually earned a doctorate degree), the easier it would be to gain the trust of those in the field—from the smoke jumpers who parachute in to fight fires to the scientists who study the conditions that create the blaze.

"As I moved through my fire career, I did everything. I did fire engines, I did dispatch, I worked on helitack (helicopter crews). I did the hot shot crews, and then the next important piece was working at the fire lab. I learned all about the research side of things, including GIS and fire modeling."

She realized how difficult the work can be and how hard it is to keep good people in those dangerous positions. She wanted to find ways to make the job of firefighters easier and safer, aiming to apply computer modeling to anticipate a fire's path and predict its intensity.

As she learned about remote sensing and computer modeling, Teske kept wondering how to put it into action. "I thought, 'Man, that's some cool stuff there, but how do you get that to the field?' As a field person, I wasn't high enough up yet to bring that kind of information back, but I knew that it was a piece that was missing. Remember, this is the early 2000s, and some of the technology wasn't there yet. Nowadays, the technology is there, and maps can be available on your phone in real time."

Using Location Intelligence to Protect Those on the Ground

"We know fires are coming," Teske said. "We know people are moving into fire-prone areas. We know that climate is impacting fire seasons. And we know that these things are all connected. So how can we be proactive and do treatments or take actions while we have conditions that give us more opportunities to control things?"

To answer this question, Teske sees technologies that can feed data into GIS playing a growing role.

"The infrared (IR) work that's out there, whether it's satellites or the IR flights at night, was a big game-changer. Instead of just going to where you were yesterday and doing what you were doing, now you have increased situational awareness, which is important for developing tactics and strategies. You can use the information from the IR in conjunction with GIS mapping and modeling, so you can see where a fire will spread the fastest and most intensely, and you can see where

treatments have happened that may reduce fire behavior or give you a tactical advantage."

Teske is excited about the use of drones for their potential to observe and stream real-time data. And she notes artificial intelligence and machine learning may increase the quality of real-time insights into the behavior of a fire.

"And then we can use the information for post-fire analysis," she said. "'Well, why did the fire do this?' Now, we can suddenly understand something about fire behavior and spread that we maybe didn't understand while it was happening. That is the power of GIS analyses and mapping."

Each insight opens the possibility of better mitigating and fighting fires while protecting ecosystems and saving lives.

Knowing where spot fires flare is important to keep fires from spreading so quickly.

Casey Teske administers a prescribed burn, one of many tactics she deploys to keep wildfires in check.

Grass and wildflowers spring up after a fire.

Hillside left barren after a wildfire.

patterns. They can see the frequency of lightning strikes, identify the number of fire-resistant trees and plants thriving in comparison with invasive species, and find historic trends for each variable.

Summing up the analysis done to help firefighters and alert the public , Teske said: "'Here's what we see. Here's our crystal ball. Here's what fires have done here in the past.' That's not going to appease everybody but helping the public understand things that are scary is going to buy us something. I just have to hope that because fire is here to stay."

"There are questions that we don't know, we haven't thought of, but somebody else might be able to answer with all this information," Teske said.

Into the Future: More Technology, Success, and Questions

Teske looks back on her own career and sees immense changes in the ways that scientists study fires and that firefighters combat them.

"Coming from the days of paper maps and drawing things on [transparent] overlays with a marker to show where the fire was yesterday and where it is today, I always say the map tells the story," Teske said. "Getting GPS units into the hands of firefighters 20 years ago was a big deal, but it changed how we were able to incorporate information. And now we have that on tablets and phones and can triangulate across cameras or use drones or whatever and stream information back in real time."

GIS enables analyzing streaming data in context. For instance, scientists and firefighters can quickly determine whether the threatened ecosystem is a forest or a prairie, and learn the relevant humidity, wind, rainfall, and snowfall

To Find People in Crisis, Illinois Rolls Out Next Generation 911

When someone in distress calls 911, the dispatcher and first responder must know where the caller is and how to reach them. The problem is, these questions can be far from clear. Cell phones don't always convey an accurate location, and street addresses don't always match the map.

"A fire department deputy chief told me they regularly get calls from fire crews they send into a neighborhood where all the streets at one intersection have the same name," said Peter Schoenfield, GIS analyst for Lake County, Illinois. "The dispatcher might say, 'Turn left on Lake Shore Drive,' but they're all Lake Shore Drive."

That scenario is typical of how addresses can become barriers to accurate wayfinding. In the US, local authorities maintain addresses and update shared maps using a GIS. Data is then shared with 911 call centers, known as public-safety answering points (PSAPs).

In Illinois, officials are working to solve this problem by undertaking a statewide modernization effort, Next Generation 911 (NG911), led by the Illinois State Police. NG911 will connect all 911 call centers and use GIS to increase address accuracy, which will improve response times across the state. NG911 has been promoted at the national level, but it's

A dispatcher handling calls for assistance relies on an accurate map to guide responders where they are needed.

up to each state to tackle the move to using internet protocol hardware and associated enhancements.

"The old systems are antiquated, particularly the call-handling equipment that telecommunicators use at each PSAP," said Cindy Barbera-Brelle, statewide 911 administrator at the Illinois State Police. "We're getting to the point where 911 systems have to look for parts on eBay, because the equipment is that old."

Modernization will unlock entirely new capabilities to better communicate the nature and location of each call for help.

"You can text and send pictures and videos to friends and family, but you can't to 911," Barbera-Brelle said. "We're really in a new environment where technology is advancing far more quickly than it ever has. Catching up to what consumers can do with their cell phones has been a driving force behind NG911."

Moving to Geospatial Routing in Illinois

Modernization will also replace the standard Master Street Address Guide—a tabular database with address ranges that ties a phone number to an address. When it was introduced decades ago, the guide was a great leap forward. It became so reliable that 911 telecommunicators could first ask landline callers, "What is your emergency?" because the location was already known. Now that most phones are mobile, telecommunicators have had to go back to also asking, "Where is your emergency?"

Illinois will answer that time-consuming and often confusing question by moving to geospatial routing and delivery of 911 calls. The new solution will automate turn-by-turn directions, using critical details in the GIS to improve location accuracy.

"Nowadays, the vast majority of calls coming into 911 centers are from cell phones," Schoenfield said. "We've been building

The Illinois Statewide 9-1-1 office consolidated multiple agencies under the auspices of the Illinois State Police.

our datasets to be able to pick up the x,y coordinate of that cell phone, drop it on a map, and pick out who the closest responders are. With NG911, a call won't even get to the dispatch center until it's been touched in some way by a GIS."

Each PSAP uses GIS to record key layers of data that include the boundaries of fire, EMS, and police response jurisdictions, street centerlines, and address points. GIS also records provisioning boundaries—coverage exceptions that get worked out between neighboring responders. "I'll cover this, even though we share it," Barbera-Brelle explained.

In addition, GIS includes many capabilities to accurately record and verify a location, represent it on a map, share address information, pinpoint a position, and direct responders to where they are needed. The use of GIS paves the way to include z-axis (vertical) support for call routing within buildings to specific floors or units.

To make the move to geospatial routing, boundaries and address data must be accurate and synchronized between PSAPs to be able to assign the right responders and route them to the right location.

"From the beginning, I knew that GIS would be at the center of deploying the statewide NG911 system that I was tasked with," Barbera-Brelle said.

Helping Each Other Achieve a Rapid Response

The legislation that put the Illinois State Police in charge of statewide NG911 includes all 911 systems in 102 counties across the state. Chicago, which is going through its own modernization effort, will connect to the statewide network. Barbera-Brelle will need to engage with 911 systems in every county and get their data to carry out the ambitious plan.

Because funds generated for PSAPs are tied to land line, wireless, and Voice over Internet Protocol (VoIP) customers, fewer land lines has meant less funding to modernize in rural areas. However, grant programs were created to provide for NG911 readiness.

To start, much of the investment went to modernizing call-handling and recording equipment to meet National Emergency Number Association (NENA) standards to deploy an Emergency Services IP network (ESInet)—a network of networks to connect all PSAPs. Funds have also gone to licenses and training for using ArcGIS Online software as a service (SaaS) and fresh aerial imagery to help counties update their maps.

"Some counties still have rural addresses that need to be updated, and I know that work is very labor-intensive," Barbera-Brelle said. "So we offered grant funding for GIS projects as well."

An additional effort involves the use of ArcGIS Hub SaaS to enhance mutual aid and collaboration. The Illinois State Police NG911 Data Hub establishes workflows for data input that walk users through each step. It increases the visibility of progress across the state with a status map that shows data submitted and data quality. GIS data technicians and administrators can access the hub to ask each other questions and share expertise.

"Once everybody submits their data, they need to work with neighbors to make sure boundaries and jurisdictions are not overlapping or gapping," Barbera-Brelle said. "GIS allows them to see where any issues are. Then they can work it out, so we have consistency."

Updating the NG911 Address Data Model

The foundation for strong address data lies in a data model within a geodatabase to standardize what is collected and how the information is stored.

"When we first saw the NG911 specs, we realized immediately that that was going to require massive changes to our road centerline datasets and the address point datasets," Schoenfield said. "We wanted an updated data model that we could put our data into, but nobody had one."

The team set out to create this foundation, aligning it with the work of the National Emergency Number Association, a key player in the move to NG911. Now NENA has a detailed NG911 GIS Data Model to describe the structure of GIS data, including field names and field data types.

"Everybody should be using that NENA spec without deviation," Schoenfield said. "And that's not just the United States. That's Canada as well, and parts of the Caribbean, [and] Central and South America. Anybody that's using a three-digit area code with a seven-digit phone number should be subject to that."

The NG911 effort has already benefitted participants who use GIS for many other purposes because it provides new levels of data accuracy and freshness.

"Having a statewide dataset that's continuously updated is something that would not have even been thought of 10 years ago," said Eric Creighton, GIS analyst for St. Charles, Illinois. "The ability for any municipality or county to be able to pull down a statewide dataset is a huge by-product of the NG911 initiative."

GIS Community Rallies to the Cause

Early in the effort, the Illinois GIS Association (ILGISA) became a gathering point for people who wanted to help improve address accuracy. Soon, they formed an NG911 advisory board, and its members have been instrumental in setting standards and workflows and checking data quality.

When streets and even street signs are under water, a map is all that responders have to show the way.

"We took the initiative, because as a statewide organization, we figured this is the perfect glove to put our hand into," said Creighton, who was president of ILGISA when the effort began. "It brought the community together. We started with an exploratory effort to understand GIS capacity across the state. We then created a schedule for the required datasets."

In the beginning, Creighton was the primary data reviewer. Now, the GIS Center at Western Illinois University has taken a larger role for data quality checks. University students participate in the work, drawing pay and gaining practical knowledge.

"We get data of varying degrees of quality," said Chad Sperry, director of the Western Illinois University (WIU) GIS Center. "We've got validation and QA/QC procedures in place that make sure that this data is standardized, it's scrubbed, and it's clean before it's ultimately uploaded to the state hub and centralized and merged. This is a really important project.

We've got double checks, and we review the data in multiple ways to make sure that we're not missing anything."

In the past, the smaller communities or GIS managers didn't have time or a strong need to prioritize data quality. "It's kind of a wake-up call for the GIS community," Schoenfield said. "The importance of our data has been elevated."

Now, the need for data accuracy is clear to everyone involved in the NG911 project.

"It's rewarding to hear first responders screaming for this data," Sperry said. "They realize they're doing a disservice to their people if they don't have good maps."

The WIU GIS Center also works more directly with 16 counties in the region that are thankful for mapping and data assistance because they have a history of not getting a lot of help.

"If you search on the word *Forgottonia*, it brings up the west central Illinois region," Sperry said. "It was coined in the late 1960s in response to the lack of transportation funding and money coming from the state and federal government. I grew up on a farm in this area, so that's where my desire to serve it comes in. We can't be left behind."

Correctly Recording Addresses

The fundamental steps of administering an address aren't difficult, but there are many ways for authorities to get addressing wrong. In Lake County, one of Illinois' more populous regions with 700,000 people in the northeast corner of the state, ongoing issues prompted a large readdressing effort.

"Back in the '70s, the county took all unincorporated addresses and standardized them," Schoenfield said. "A handful of subdivisions complained, and the county acquiesced and let them keep their old address numbers while maintaining five-digit addresses for county government use. A few years ago, we realized this 'dual addressing' was a problem that needed to be fixed, so we crafted a new address ordinance that allowed those nonstandard addresses to remain if they were logically consistent—even on one side, odd on

the other, with addressing flowing from low to high. When this couldn't be maintained, new addresses or even street names were issued to correct the problem."

The move was essential to create the consistency a computer can understand and that logically flows along a street, but it was far from trivial.

"It was a major chore," Schoenfield said. "We had to meet with each homeowners' association. We had to send letters out to all of the different stakeholders, not just fire and police, but sheriff, and all the different utility companies, the county clerk, the treasurer. We also had to talk with the data supplier of delivery services like FedEx and UPS to make sure they got our changes as fast as possible. Anybody that had a stake in the address, we had to notify that the address was changing."

For delivery, the post office is the top priority.

"You have to get the post office the right address, or it's not going anywhere," Schoenfield said. "We send over a list, and then check it every month to make sure they are recorded correctly. If we have any fixes, we wait another month and check it again. It's a very involved process, and you really have to have the fortitude before you embark on something like this."

For first responders, changes in address flow cause problems that Schoenfield and his team hear about often.

"Each fire district we consulted with would tell us about streets that were messed up," Schoenfield said. "Addresses might start with three digits in a municipality, but then when you get halfway down the street, they jump. Crews that don't know the area turn around, thinking they missed the address without realizing they have to go another mile before the address system changes again."

When time is critical, even momentary confusion can have dire consequences.

"First responders will tell you that two minutes is the difference between life and death," Schoenfield said. "That's why we need to make these changes. When NG911 is fully operational, we'll all start reaping the benefits."

How Technology and GIS Students Aided Response to the Great Flood of 2019

It's called the Great Flood of 2019 for good reason. It was the wettest spring on US record, impacting 14 million people as multiple storms hit and rivers overflowed, flooding the Midwest, the High Plains, and the South from January through June. New high-water marks were set in 42 locations along the Mississippi River. The lessons learned then continue to this day—the flood was one for the record books across many areas of science.

Students at Western Illinois University (WIU) in Macomb had a front-row seat—Macomb sits between the Mississippi and Illinois Rivers and north of where the rivers meet—and they set to work monitoring impacts and identifying areas of need.

"The Mississippi River was having extreme flooding, and the Illinois River just couldn't drain," said Chad Sperry, director of the GIS Center at WIU and a member of the state incident management team.

Sperry was dispatched by the Illinois Emergency Management Agency (IEMA) to the State Unified Area Command (SUAC) in Winchester. He brought a team of GIS students to create maps that could help first responders from IEMA, the National Guard, Illinois Department of Transportation, Illinois State Police, US Army Corps of Engineers, and others.

"There were a lot of road closures, so the students were involved in making detour route maps and other mapping products," Sperry said. "We used drones to provide a real-time situational awareness capability. We were working 14-hour days for 16 days straight."

In many of the small communities along the rivers, flooding is a regular occurrence, but not to the extent and duration of these events. Beardstown, Illinois, experienced 176 days of minor and moderate flooding. In nearby Havana, major flooding stretched for 37 days.

"The duration and the intense months of coping with flood response were exhausting," Sperry said.

Providing a View of an Important Levee

In the state's unincorporated community of Nutwood, Sperry and his team monitored the main stem levee system a few miles from town.

"Nutwood isn't significant in terms of population, but very significant in terms of impact," Sperry said. "It sits right at the bluff, so it's almost out of the floodplain but not quite. It became evident that this levee was going to fail due to the models that were run by the US Army Corps of Engineers and the National Weather Service using hydraulic forecasts and GIS to do predictive modeling."

The townspeople built their own backup levee using bulldozers to push dirt from the farm fields around town. Still, the main levee failed, and then the town levee failed. When the Nutwood Levee overtopped, it forced the closure of Illinois State Route 16 at the Joe Page Bridge near Hardin. And it took weeks for the waters to recede.

Sperry and his team were there over the course of the levee failure.

The drone operator (above) provided a helpful perspective to keep an eye on the temporary levee in Nutwood (below).

"We mapped Nutwood using drones and [ArcGIS®] Drone2Map® technology before the levee breach, during the levee breach, and after," Sperry said. "Detailed drone mapping and elevation models were used by IEMA to inform evacuations."

The team also used GIS tools to record the aftermath. They used ArcGIS Survey123 and ArcGIS Collector to do post-flood damage assessments, photographing everything. Collecting data on iPads, the team moved away from the paper-based system that had been used historically up and down the Illinois and Mississippi Rivers.

Sperry helped train everyone and keep them on task. In some cases, this required guiding people through the discomfort of new work processes.

"When you have folks used to writing everything down on paper, the first little hiccup makes them want to ditch the iPad," Sperry said. "We heard some of that. But when we got to lunch and hooked up all the iPads to a hot spot and pushed data to the cloud so they could all see their work on the map, the light bulbs started going on. It was one of those aha moments for a lot of damage assessment crews that had never used this technology."

Keeping Rural Communities Mobile

The work to map transportation routes was one of the more critical elements, with floods closing many roads and bridges and people needing to evacuate.

"The Department of Transportation had an expert traffic flow modeler that was optimizing evacuation routes," Sperry said. "We imported that information into geodatabases to take it to the next step, creating map products to provide the context of where all the people would ultimately end up."

Many ferries cross the rivers because of a lack of bridge infrastructure, and road and bridge closures from the flooding

Sandbags and boards placed on top of this levee helped save the farmer's field. (Photo courtesy of Chad Sperry)

made mobility even worse. Commuters to Saint Louis, Missouri, even used their own boats to make the crossing, parking their cars across the river to avoid having to spend three additional hours each way driving around the flooding.

As the high water moved downriver over time, the team did inundation modeling to help understand impacts as flooding neared Saint Louis.

"There were some questions that if a particular levee breaks, what would we be looking at?" Sperry said. "We were creating flood extent maps to examine the realities if any one of these levees breached."

Incident commanders and planning section chiefs studied the flood extent maps to create contingency plans. By looking at potential outcomes, they could determine which homes and roads would be impacted and prioritize evacuation areas should the levees fail.

"We used something like one million sandbags during the event," Sperry said. "Levees are typically built with an earthen core with sand over the top of it. Over time, the sand and core get saturated, and that puts pressure and stress on the surrounding soils. We had boils popping up a half mile inside the levee where the river found a path, and they would put sandbags around those to equalize them with the height of the river."

The US Army Corps of Engineers used ArcGIS Collector to mark and monitor the boils, any depressions in the levee, and anything out of the ordinary. That data about weak spots will then be used to inform future levee improvements.

One View for the Team of Teams

The real-time data collected by different agencies and GIS students was fed into ArcGIS Dashboards and shared across the state.

"We built a dashboard with the National Guard to show where sandbag troops were being deployed," Sperry said. "After that first briefing where we used the dashboard, they moved us into the main building, and the dashboard stayed up on the main screen."

Soon the dashboard was shared with the Emergency Operations Center in Springfield, Missouri, and the National Weather Service in Chicago used it to see what was really going on from a levee status standpoint. It was used to brief the governor, department heads, state senators, and US senators.

Eventually, the dashboard aggregated and consolidated data from 10 to 15 GIS analysts working for various agencies. WIU students worked alongside the experts. Pam Brooks, GIS specialist at IEMA, had already fostered relationships with GIS technicians from other agencies and was able to help coordinate the collaboration.

"Whether there's a need for specific map elements to be created or various datasets to be updated, we all chip in," Sperry said.

At daily morning briefings, the teams discussed any status changes on the dashboard. For instance, the representatives from the US Army Corps of Engineers would inform the group of any levee breaches or overtops or levees in a state of caution.

The students kept track of details such as shelter locations because sometimes shelters would have to move if a levee failed. Keeping that information up-to-date was crucial to making sure evacuees had somewhere to go.

"We would get requests to add something, such as weather overlays for radar, and the students would research and find the best live data to add to the dashboard," Sperry said. "There were many hands-on opportunities."

Lifelong Lessons Learned

The students gained a tremendous learning experience from these events, with immersion in the use of a wide variety of GIS tools and the need to deliver answers quickly during a crisis.

"We got a call one night just as we were getting ready to go home for the evening that a levee had just overtopped," Sperry said. "So everybody just set their bags down and dug back in again. We were there for a couple more hours that evening."

The flood events gave students crucial practice in the fast-paced, high-stakes world of emergency response using GIS, a common and important application of the technology.

"It was definitely the most stressful work environment I've ever had to work in," said Ian Stearns, a WIU student majoring in meteorology who helped out. "Being able to learn how to control the stress of all the things going on, all the decisions you have to make, has been really helpful for me."

At the time he joined the student team to respond to the disaster, Stearns had taken one GIS class and was working at the GIS Center for three months in a paid position that gives students real-world experience.

"When we flew over the temporary levee in Nutwood to identify places it might fail, that was really fascinating," Stearns said. "The way we were able to create a detailed digital elevation model from the imagery—and Chad Sperry was able to model water height in relation to it and other buildings—was awesome. I had never even thought of that kind of application of GIS."

In between events, the students would talk about GIS jobs and get to know the emergency personnel. The students' real-time skills made an impression on their professional cohorts.

"One of our grad students had two different job offers from the US Army Corps of Engineers before he even got home," Sperry said.

Today, in Stearns' meteorology classes, he said students talk about the accumulated snowfall, and how the ground was still frozen when the snow was melting so there was no intake of moisture in the soil, and then there were above-average spring showers. So all those factors contributed to why this event was so historic.

Though the flood of 2019 set many new records—as longer in duration with higher floodwaters—Sperry said the damages were less than what was anticipated.

"We had eyes in the sky and the ability to predict and not just react," he said. "Instead of waking up to find that a levee broke in the night, we deployed sandbagging efforts to where GIS predicted it would break. We knew what was coming with the rainfall models and the gauge models. And so, technology was really given a lot of credit for minimizing the impacts."

Defense and National Security

Analysts at the National Geospatial-Intelligence Agency (NGA) often work behind the scenes, providing maps and data to policy makers, the armed forces, intelligence agencies, and first responders. Their work—much of it done using GIS technology—supports critical military, political, and humanitarian missions worldwide.

Recently, NGA analysis teams have adopted artificial intelligence (AI) and machine learning to process data and help speed the delivery of powerful geospatial intelligence, known as GEOINT.

There's no doubt that computers excel at processing lots of data, connecting dots, and uncovering trends and patterns. Machines process data faster, and humans can get overwhelmed by large volumes of data. However, computers are not creative thinkers. They don't work collaboratively with others to bring diverse viewpoints and opinions together, and they can't gather around a map to discuss and debate the benefits of different strategies or formulate the best path forward through consensus.

When you pair humans and computers together you get a powerful team to take advantage of the explosion of available data. Remotely sensed imagery has grown exponentially, and sensors and sources that provide spatially and temporally referenced data are everywhere you look.

Defense organizations talk about delivering GEOINT at the "speed of need." All static maps are old quickly, and the best map is the one that was just created. With access to many dynamic layers of data, defense mapping professionals can use the latest knowledge and deliver information that fits users' needs.

When every second counts for life-and-death decisions, quick and clear communication is critical. The US military has developed many methodologies over the years to aid clearer communication, such as the 9 Line report that is used to communicate unexploded ordnance, medical evacuations, and other critical information.

The 9 Line goes all the way back to radio reporting. Line 1 communicates the location, line 2 reports how the reporter can be contacted, line 3 shares the details about the scope (such as the number of injured), and onward until all critical details are conveyed. Orienting both the reporter and the receiver to the nine lines on the report helps both parties understand what's being sent and what's being received.

The 9 Line format is now used in GIS apps to report threat inspection details on a survey form. The common operational picture provided by GIS helps synchronize perspectives between people in the operations center and the field. By sharing information via a formcentric app, users can speed up the transmission of inspection details for chemical, biological, radiological, or explosive hazards. The details are sent directly to a dashboard and quickly aggregated in a map view.

Making quick, prudent tactical decisions requires an improved understanding of what's happening now. With real-time information delivery and rapid analysis, defense GIS users can optimize limited resources to support the mission.

US Navy Puts Tactical Tools to the Test

The US Navy's annual Innovation Battle Lab brings together more than 700 people from special operations, intelligence, and law enforcement to test and try the latest technologies in live combat-simulation scenarios. Over the course of two weeks, participants run more than 250 experiments. The event is unique in how it addresses the needs of operational activities in a flexible way.

Although the event has many predefined scenarios, it's mostly about bringing together operators, analysts, and technicians to experiment with new methodology. Together, these teams validate new gear and technologies—judging the effectiveness and applicability to operations.

The idea is to create a canvas where capabilities and people with great ideas can come together and craft new strategies.

The event mixes and matches technologies and outside technologists with "experiment to failure" as the mantra.

"They really try to push the limits," said Jim Bennett, an Esri expert on naval special operations. "Push it until you break it. If you break it, let's fix it and keep pushing it."

The constant problem solving, exposure to leading-edge technologies, and mission-driven testing that clearly marks progress against objectives make for eager participants at the event. The days go quickly, and the cycle of compiling lessons learned and preparing for the next Innovation Battle Lab continues throughout the year.

Drones Capture Basemap Imagery

Military adages frame each operation. The five Ps, for example, remind teams that "proper preparation prevents poor performance." In the collaborative and fast-paced environment of the Innovation Battle Lab, participants vet technology against such truisms, looking for better ways to collect, analyze, and share information in rapidly changing scenarios.

The warfighter needs to know where they are and where the enemy is. Being able to get the right data to the right place at the right time is the bottom line.

Understanding the "where" component requires an up-to-date map. To achieve this, participants set a pre-event priority of collecting high-resolution imagery and elevation data using a professional-grade drone to create a basemap within a GIS.

"We collected the drone data like a surveyor collects their data," said William Shuart, GIS coordinator, drone operator, and professor at Virginia Commonwealth University. "This involves setting up in the right location, using a [global positioning system] GPS unit to log precise location data, setting up ground control points, making flight plans with overlap, and collecting vantage points of objects from at least two different angles."

Collecting imagery over a 2,000-acre target area took a full day, five flights, and resulted in 7,000 captured images for the latest lab.

US Navy SEALs practice a beach landing during the combat-simulation exercise.

"Drones have a specific job, to collect imagery," Shuart said. "Because they're so advanced, it allows us to get a really good picture of an entire landscape really quickly."

The imagery was then processed using ArcGIS Drone2Map, which automates the production of 2D and 3D imagery products to fuel rapid analysis and visualization. Specialized workflows guided the review of imagery captured in each flight.

"If you did all that stuff right and you have good GPS data, then you can get really good spatial accuracy," Shuart said. "That's really the goal, because we're going to feed the orthophoto, the photo point cloud, and the 3D mesh into a GIS to bring it together with other data, create web scenes, and do advanced analytics."

Sensors Fill Information Gaps

Sensors can be used to fill intelligence gaps—the known unknowns. The Innovation Battle Lab makes use of airborne and waterborne sensors that capture imagery or use radar and other wavelengths that the human eye cannot detect.

A growing class of sensors are unattended ground sensors, which employ sensing modes such as seismic, acoustic, magnetic, and infrared to detect the presence of people or vehicles. These sensors detect and relay signals via radio frequency or satellite communications to personnel in the field and to a central command. Often the signals are displayed on a map to give geographic context.

Many on-the-ground sensors are deployed during the Innovation Battle Lab. To assist in tracking and managing these valuable and sometimes pricey assets, the team found a way to track the location and owner of each sensor.

"We set up a form to collect contact details and plot sensor placement on the map," said Bennett. "Each sensor gets a bar code that can be scanned to reveal contact details. It also set in place a system to notify when and where sensors are deployed and recovered."

Easing Technology Integration

Another high-priority pre-event activity involved laying the groundwork to get all sensors and other technology on track.

An integrated sensor architecture modernizes traditional stove-piped integration approaches. The design, based on the principles of service-oriented and open architectures, addresses integration without requiring users to customize interfaces for specific sensors or systems.

The resulting data layer pulls together all sensor feeds into one service for easy integration within GIS and other technologies. The result is a plug-and-play design to support interoperability across all the technologies.

Although details of the scenarios are classified, the spirit of the event is open. Because of the advance work to integrate technologies from more than 50 participating companies, teams are able to test and try interoperable solutions.

The goal is for a capability such as ArcGIS to tap off the sensor data spigot and get access to everything without having to go after dedicated point-to-point integration with every sensor. Participants at the exercise comment that there's barely enough time to shake everybody's hand, let alone build interfaces to everyone's technology.

US Navy SEALs rappel from a helicopter onto the roof of a building during the training exercise.

The long-range, high-altitude Lockheed SR-71 Blackbird reconnaissance aircraft flew so fast that it could outrun missiles.

Cold War Spy Imagery Now Helping Fight Climate Change

Throughout the 1960s and 1970s, a top-secret US program dubbed Corona by the Central Intelligence Agency (CIA) captured satellite images of Soviet military installations. Now, the more than 850,000 photographs can help scientists analyze human impact on Earth through the decades—changes such as tree loss and urban sprawl.

Declassified in the 1990s, the massive imagery collection remains mostly disorganized and stored on film. Each image must be developed and manually identified with its exact place and time—an arduous and expensive process. Recent technology advances are giving the Corona project new purpose. Researchers are digitally correcting the sometimes blurry images and using GIS technology to orient the images in space and time, at scale.

Old Images and New Technology Solve Problems

The more data that scientists can collect and analyze from the past, the more accurately they can model what will happen in the future. In the 1960s, just as the Corona project was launching, early concepts of quantitative and computational geography emerged. Decision-makers quickly grasped the power of these approaches to solve problems, and the first GIS launched in 1963. As Corona satellites began circling the earth, Esri was getting its start with applied computer mapping and spatial analysis. The tools, first created to help land-use planners and land resource managers make informed decisions, were eventually released as software.

Today, GIS endures as an enterprise technology uniquely able to analyze data from nature, society, and business. It's capable of taking enormous amounts of information from different sources and distilling it as comprehensible and actionable for nearly anyone. GIS is as useful in understanding water scarcity as it is in optimizing business operations via a digital twin.

One team in Saudi Arabia is using GIS and satellite remote sensing methods to study shoreline changes in Yanbu, the world's third-largest oil refinery. The location intelligence gathered from their study is allowing the team to understand how human development has impacted the shoreline since 1965—an understanding that wouldn't be possible without Corona imagery. They'll use that data to predict the continued threat to natural coastal systems and project the consequences of that damage .

Another group of researchers is studying Nepal's second-largest lake, Phewa Tal or Phewa Lake. Scientists have long believed that the lake is shrinking but it's been disputed because of a lack of historical data in the region. Team members analyzed old maps and Corona satellite imagery using GIS and AI to understand lake shrinkage over time. Their analysis revealed that landslide activity, sedimentation via the watershed, and urban development will cause the lake to lose 80 percent of its storage capacity over the next several generations if left unaddressed.

Evolution of Imagery Technology

During its tenure, the Corona project yielded nearly a million images. The first satellite alone returned more imagery of the Soviet Union than the entire Lockheed U-2 high-altitude reconnaissance aircraft program. At the time, there was no ability to digitize these images or put them in a database.

Now GIS tools allow researchers to display many images and catalog and compare them. Users can navigate through multiple layers of imagery to look through space and time. This visual comparison is just a start, with great value derived from imagery analysis that goes beyond what the eye can see.

Imagery is cell-based, with each pixel containing values that the GIS can classify for layers of information such as land or vegetation type. Using a geoprocessing model, datasets can be derived directly from imagery. Most imagery analysis work is now automated, using online computing capacity to process and store a daily refresh of new images provided by today's satellites. Applying the AI method of machine learning allows computers to detect objects and pull out such elements as building footprints or missile launching sites.

By late 1960, data derived from Corona imagery allowed the director of the national reconnaissance program to declare confidently that the Soviet Union did not possess more strategic nuclear weapons than the US. Reaching that conclusion required many more human hours than would be necessary with today's technology.

Before the release of the Corona images, satellite imagery only stretched back to 1972 when the first Landsat satellite began transmitting digital images of the earth from space. Despite being static and black and white, the Corona images are of much higher resolution than their Landsat counterparts because they were taken closer to the surface and at the optimal time of day for discerning objects on the earth's surface.

Extending the Historical Timeline to Model the Future

Researchers rely on historical information collected from books, photographs, journals, artwork, and other sources to help extend the timeline of understanding. Projecting that knowledge into the future, they use imagery with GIS analysis to assess risk and vulnerabilities across a wide range of industries and areas of study. To gauge systemic threats such as climate change and sea level rise requires a long view—using sometimes at least 30 years of data.

For instance, scientists are using detailed logs from whaling vessels to understand what climate and marine conditions were like in less-traversed areas of the ocean in the 19th century. Others venture into caves to collect samples from bat guano for an accurate snapshot of the climate in that area over time. Photos taken by Cold War spy satellites fit into this category as well.

One of the first people to recognize the potential usefulness of the Corona images in studying our planet was former Vice President Al Gore. In 1990, the then senator Gore wrote a letter to the CIA asking if Corona satellite imagery could be used to help solve environmental mysteries. The resultant MEDEA program, led by 70 scientists, became one of the greatest sources of earth science data in the world. Information from MEDEA's Global Fiducial Data Network assists in monitoring critical environmental parameters today.

As technology advances, researchers are finding new ways to access incredible volumes of data. For bigger issues such as climate change, they need information that spans across time. When scientists combine Corona images with GIS, they can see changes through the decades—insight crucial to making more informed decisions to protect the planet.

The Fairchild C-199 Flying Boxcar snagged Corona satellite film canisters as they fell from space.

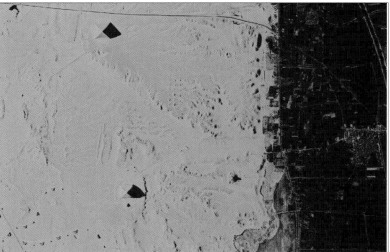

Corona images provide important details about how the world looked in the 1960s and 1970s in *(clockwise from upper left)* Baghdad, Iraq; Egypt; and Moscow.

(US Army National Guard photo by Staff Sgt. Josiah Pugh)

In Back-to-Back Hurricanes, Louisiana National Guard Calls on Predictive Model

Hurricane Delta made landfall in Louisiana in mid-October 2020, resulting in two casualties, flooding neighborhoods, and leaving thousands of residents without electricity. The storm hit just six weeks after Hurricane Laura devastated the same region, and just as power had been restored. With back-to-back hurricanes, the state's emergency responders called on new models and maps to anticipate where help would be needed most.

Hurricane Delta—the third to hit the state and the fourth major storm of the year—left a deluge of more than 15 inches of rain that damaged homes and businesses already reeling from Laura's winds, which tied an 1856 storm for the strongest gusts to hit the state.

With a hurricane bearing down, Colonel Greg St. Romain, commander of Louisiana National Guard's 225th Engineer

Brigade, used maps informed by a new predictive model to guide his team into position.

"The storm surge map helped me preposition the right types of equipment for a windstorm versus a flooding event," Col. St. Romain said. "We have high-water vehicles, we have boat systems, and we have engineering equipment to clear roads so emergency vehicles and power companies can access impacted areas."

The maps, showing where wind would cause the greatest damage, were developed using models built with a GIS. During Hurricane Laura, the standard storm surge model showed significant flooding for areas near Lake Charles. However, as the hurricane approached land on August 28, 2020, a high-resolution storm surge model coupled with a consequence model revealed high winds as the greatest threat.

"Time is of the essence, and having the right tools and the right data is essential to make effective personnel and equipment decisions," Col. St. Romain said.

Arriving at a Predictive Model

The consequence model was developed at Louisiana State University (LSU), born from the need to bridge the gap between a high-level storm surge model and an understanding of local impacts. It uses the high-resolution Advance Circulation (ADCIRC) model of storm surge provided by a team of researchers at the University of North Carolina and Notre Dame University. The model is visualized through the Coastal Emergency Risk Assessment software to predict impacts on an inventory of infrastructure assets, businesses, buildings, homes, and people. Together, these models have proven their predictive power.

"When I first came here, Hurricane Isaac had just impacted the state [August 21, 2012], causing massive flooding in LaPlace, which previously hadn't experienced any significant

flooding from storm surge," said Brant Mitchell, director of the Stephenson Disaster Management Institute (SDMI) at LSU. "We went back and looked at the data and analyzed the ADCIRC outputs, and it successfully predicted approximately 90 percent of the extent of flooding 48 hours prior to landfall. The model wasn't leveraged by the emergency management community then, which would have helped the response. So, we set out to provide the output from the model in a manner that can be easily understood."

Emergency management professionals have trouble reviewing complex models during lifesaving and property-saving measures. They're busy allocating resources and triaging calls for assistance. Mitchell realized that emergency managers and public officials needed a way to quickly grasp the information in the models.

"Decision-makers need tools that tell them exactly what's going on," Mitchell said.

The National Weather Service (NWS) model is fast and can be run on a desktop computer. Speed is crucial to understand

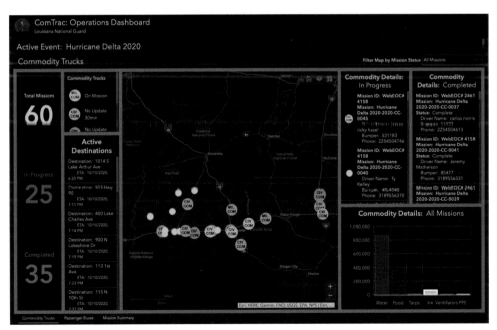

ComTrac provides a live view of the status of supply missions, with real-time vehicle tracking to show the location of each delivery truck. The app is built using ArcGIS Dashboards, which ingests GPS signals from phones to provide an automatic vehicle locator.

Louisiana National Guard members help clear roadways and assess damages on August 27, 2020, in Lake Charles, Louisiana, after Hurricane Laura made landfall the night before. (Courtesy of the Louisiana National Guard, photo by Staff Sgt. Josiah Pugh)

the potential impacts while meteorologists are developing the hurricane advisory. The NWS's Sea, Lake, and Overland Surges from Hurricanes (SLOSH) model doesn't have the granularity or resolution of the ADCIRC model that researchers run on supercomputers or the data inputs of the consequence model. "Even with a supercomputer, it takes approximately 90 to 120 minutes to run the model," Michell said. "That doesn't provide the National Weather Service sufficient time to integrate it into the development of their advisory."

The more granular model fits perfectly with the emergency response mission. To make the details easy to consume and share, Mitchell and the team from SDMI deliver detailed maps via ArcGIS StoryMaps, share the data, and then print large maps to convey actionable details.

Models Inform Operations

The SDMI consequence model pinpoints a hurricane's impact on people, homes, businesses, hospitals, and critical infrastructure—details of great importance to first responders.

"The model presents an overview of damages in a manner that is easy to understand," Mitchell said. "It serves as a decision-making tool on whether to call for evacuations and where to prioritize post-disaster actions such as search and rescue."

In the case of Hurricane Laura, the ADCIRC and consequence model saved needless work when it showed that storm surge wasn't likely inside the city of Lake Charles. With Hurricane Laura designated a Category 4 hurricane, the National Weather Service warned residents of Louisiana of the potential of up to 20 feet of flooding and life-threatening storm surge.

"We have an armory in Lake Charles, which is used as a staging area, and they were talking about relocating those resources," said Mike Liotta, GIS manager and a civilian with Louisiana's Military Department. "After looking at the consequence model, I was able to update our leadership that our staging area should not be impacted by the storm surge."

Having the maps and models of the projected damage also helped troops stage the proper equipment for cleanup and to stand down the search and rescue mission. It typically takes 24 hours to completely switch missions, but thanks to the predictive view, the 225th Engineer Brigade was ready.

Maps Support Impact Areas

GIS outputs take the form of a real-time common operational picture (COP) within an emergency operations center (EOC), a place where both Mitchell and Liotta can be found during an emergency. Staff with phones or tablets in the field use apps to collect and sync much of the data that a COP displays. The team also uses GIS to plot outputs of the high-resolution model onto large maps for boots-on-the ground first responders who often work in off the grid areas without communication signals.

"I like having something printed in my hand when I leave the office and head to an impact area," Col. St. Romain said. "I can lay it on the hood of a vehicle and really study it and understand where I need things to be. I use it as a kind of sand table to maneuver our equipment and personnel."

Huddling around the paper map with a pen in hand

continues to be a great way to collaborate with a shared context. However, it's never just one map and done.

Col. St. Romain was an engineer officer within the ranks for many years, before taking command in May 2021. "I've had opportunities to work with our GIS department in the past," Col. St. Romain said. "I know what Mike Liotta and his team are capable of delivering. If I'm not comfortable with the incoming information, I reach out to him to reinforce a decision or provide me with more details of an area I'm concerned about."

Taking Care of Consequences

After Hurricane Laura left considerable damage, Col. St. Romain shifted his focus to a long-lasting and far-reaching debris removal exercise, and again relied on GIS.

"If I had an available vehicle, we were utilizing it throughout the event," Col. St. Romain said. "Our operations spanned from the southwest coast of Louisiana to the northern parishes."

The team from SDMI deployed a real-time truck-tracking dashboard at the state EOC to inform and reassure leaders that all equipment was where it needed to be. Mitchell,

The Louisiana National Guard had more than 6,200 Guardsmen supporting current and future operations following Hurricane Laura's landfall, including high-water vehicle evacuations, commodities distribution, route clearance and road debris removal, and infrastructure inspection support in southwest Louisiana on August 27, 2020. (US Army National Guard photo by Staff Sgt. Josiah Pugh)

previously a member of the National Guard and currently a lieutenant colonel in the Army Reserves, designed what's known as the ComTrac system to track resources as they move after a major disaster.

"The National Guard moves commodities, whether it be food, water, tarps, and other vital resources," Mitchell said. "ComTrac gives everyone the ability to identify where each truck is located once it leaves the regional staging area."

The high-resolution confidence model and the ability to track resources in real time have given the Louisiana National Guard new inputs for better understanding.

"It's never a guessing game," Col. St. Romain said. "We rely on best available tools to make the most educated decisions. The consequence model definitely reinforced our decisions to readjust and pivot to certain parishes."

Guardsmen stored boats and shifted to debris pickup when the model showed that storm surge would be minimal in Lake Charles. (Courtesy of the Louisiana National Guard, photo by Staff Sgt. Josiah Pugh)

Hollywood Tricks and Location Technology Catch Poachers of Sea Turtle Eggs

On a warm Saturday night in 2018, somewhere in northwest Costa Rica's Guanacaste Province, Helen Pheasey inspected a small white spherical object in the palm of her hand. Pocketing it, she took a short walk to a nearby beach to begin the night's research.

It didn't take her long to find an olive ridley sea turtle, measuring 72 cm tip to tip, making its slow progress out of the surf and onto the sand. Pheasey watched as the turtle dug an egg chamber with its rear flippers and settled itself above it. Over the next 20 minutes, it lay a clutch of around 100 eggs, the same color and shape as Pheasey's spherical object.

When she knew the turtle was roughly halfway through its task, Pheasey reached down and deposited the object among the eggs. Then she made her way home to wait.

On Monday morning, the object, which had an embedded

GPS transmitter, began to move. It was heading away from the beach, deeper into Costa Rica's Central Valley.

On the Poachers' Path

GPS-enabled tracking systems represent one of the largest segments of the GPS industry. Trackers generate $1.7 billion in annual sales, a figure expected to double over the next seven years.

Every tracking system has three basic components: a GPS receiver to plot location; a communication device, such as a cellular connection, to transmit those location coordinates; and a GIS to transfer positions into tracks on a map.

Most of these systems are used to capture the movement of people—usually by companies or government organizations interested in the efficiency of fleets, for instance—but some are used to track animals.

Pheasey, a conservation biologist and doctoral candidate at the University of Kent's Durrell Institute of Conservation and Ecology in the United Kingdom, uses a tracking system to monitor two sea turtle species—the green sea turtle and the olive ridley— that have seen a sharp decrease in global population. An estimated 800,000 olive ridley females nest each year, down from a historical estimate of 10 million before overexploitation for meat, eggs, and leather. As part of her analysis of Costa Rica's black market in sea turtle eggs, Pheasey fools poachers into picking up the fake eggs.

The decoy, called an "InvestEGGator," was developed by conservation group Paso Pacífico. Kim Williams-Guillén, the group's lead conservation scientist, drew inspiration from an unlikely source. On the TV series *Breaking Bad*, drug enforcement agents attach a tracking device to the underside of a barrel of chemicals used to manufacture methamphetamine, hoping the cargo's path will provide useful information about the drug supply chain.

Williams-Guillén collaborated with Lauren Wilde, a makeup artist who works with cinematic special effects teams. After studying the composition of sea turtle eggs, Wilde used a polyurethane filament for the housing. A 3D printer creates half an egg. After the electronics are inserted, the second half is printed and fused to the first.

Pheasey learned about the decoy eggs as she was beginning her work in the field in Costa Rica. "The two projects were so compatible," she said. "I'm looking at the illegal wildlife trade, and they're looking for someone to put these things in the ground. They had a prototype ready to scale but no place to deploy it."

Will It Fool the Pros?

Over a period of several weeks, Pheasey and her team deployed 101 decoys. They began on the country's Caribbean coast, where the larger green sea turtle is more common, before moving to the Pacific side to study the olive ridley.

The team's presence didn't draw much suspicion from poachers, who are accustomed to seeing turtle researchers on the beaches at night. Even the deployment of the decoy eggs looks unsuspecting since researchers often reach into the nests to deposit humidity monitors and other sensors.

Nor did the poachers pose much danger to the scientists. Because the payout for this kind of poaching is relatively

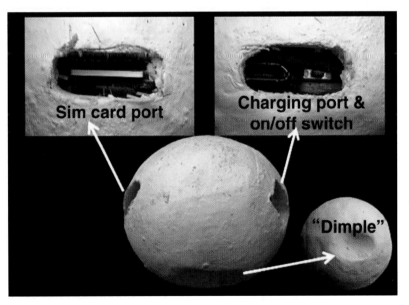

The realistic egg exterior hid sophisticated electronics.

An olive ridley sea turtle digs a hole in the sand to lay eggs.

A Glimpse of the Supply Chain

On a GIS-enabled phone app, Pheasey could follow the egg's progress away from the beach. "It just went on and on," she recalled. After traveling 137 km, it remained stationary. Pheasey zoomed in and pinpointed the location to an alley behind a supermarket.

"That back-alley transaction was probably with someone who was then going to sell them door to door," she said. "So you had the whole trade chain and a good indication of the number of players."

With the concept successfully proven, Pheasey hopes the decoy eggs reveal even more details about the larger network. Rather than battle individual poachers, authorities could have a bigger impact by targeting the overall supply chain.

"We want to have enough projects in different locations so we can identify problems and refine

Sea turtles are far more at home in the water than on land.

small—a dozen eggs sold to a trafficker will net about $4—the researcher's activities don't put a huge dent in the income of individual poachers. "It's quick and easy money for them," Pheasey said. The hard-boiled eggs are a popular bar snack. They are also consumed raw in a cocktail called sangrita or used in a dish that's similar to tortilla de patatas, or Spanish omelet.

Pheasey knew many of the decoys would not yield useful information. Poachers sometimes spotted them.

Still, the results were promising. Of the 101 fake eggs, 25 were taken by poachers. Five provided trackable signals. Most of the paths ended close to the beach. These were of limited interest. Pheasey was less interested in small-time poachers than in the traffickers that deal in bulk. And one Monday morning, she found what she was looking for.

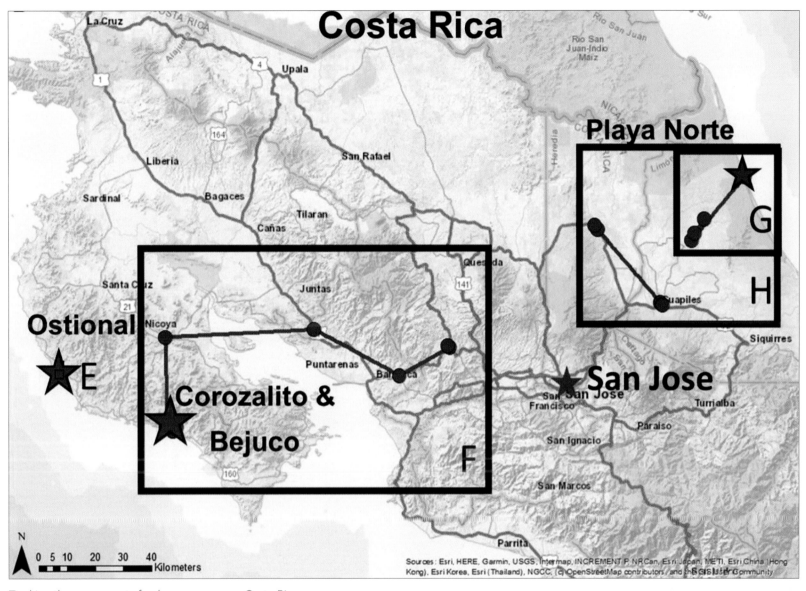

Tracking the movement of a decoy egg across Costa Rica.

the technology so that it can be a usable tool for law enforcement," she said. "But for now, it's about identifying traffickers and those who are moving large quantities, rather than individuals who are probably from marginalized backgrounds. We're not interested in taking the pittance the poacher is getting. We're more interested in enforcing the law on a larger scale and waiting behind that supermarket loading bay for the next shipment to come in."

Protecting more animals could also be in the mix with

Pheasey's research having the potential to inspire poaching mitigation efforts for other species and in other locales. Paso Pacífico is even considering how to apply the tracking concept to battle the large black market for hammerhead shark fins.

"The most comparable species to turtles is crocodiles, because in some countries they're poached for their eggs," Pheasey said, while noting that depositing eggs under a mother crocodile poses some logistical challenges. "It would definitely require a different deployment strategy."

Cybersecurity: The Geospatial Edge

When Russia annexed the Crimea region of Ukraine in February and March 2014, it shocked the world and was a surprise even to many experienced analysts and organizations monitoring Russian activity. But a few experts saw indications and telltale signs beforehand.

Volker Kozok, a lieutenant colonel and cybersecurity expert in Germany's armed forces, was one of those few. He tracks digital security threats and devises countermeasures. The job gives Kozok a front-row view into the current age of hybrid warfare.

Today's military conflicts rarely play out only on physical battlefields. They are just as likely to include cyber attacks; assaults on critical infrastructure; and various forms of weaponized information, intended to sow seeds of confusion and insecurity in a population.

In the case of the Russian invasion of Ukraine, what Kozok and his colleagues noticed in January 2014 was a clandestine opening salvo from Russia. They observed that Russia was installing what looked like an undersea cable across the Strait of Kerch, a narrow waterway between the Black Sea and the Sea of Azov. As they monitored progress on the 46 km cable, it became apparent that its end point was the peninsula of Crimea.

The cable's existence strongly suggested that Russia was making a move to connect Ukraine's critical infrastructure

with Russia's. Moreover, it looked as if Russia's cable would carry high-speed internet communications that could bypass Ukrainian service providers.

The Team Plots the Progress of Hybrid War

Kozok's team members bolstered that inference through the sophisticated use of GIS technology. They could examine the cable's construction in the context of location on a map that showed the world's undersea cables and the nodes that connect the internet. From this, the team could deduce that Russia was aiming to control online communications. In the annals of warfare, it was a defining moment.

"It was the first time a country had organized a military attack while also being very smart about planning for connectivity using a sea cable," Kozok explained. "Someone had to drive to Crimea as a tourist [before the invasion] and figure out the

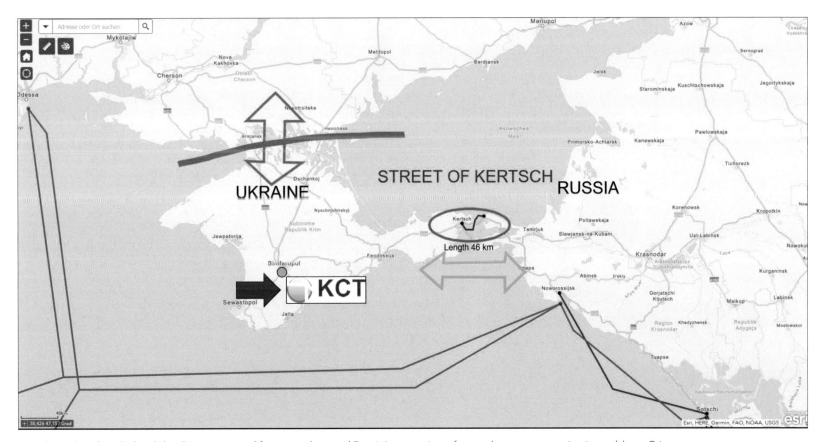

Map-based analysis helped the German armed forces understand Russia's extension of an undersea communication cable to Crimea.

cable's entry point. They had to make those plans without actually controlling the country."

Russia continued to employ hybrid war techniques in Ukraine. Russian hackers launched a cyber attack against Ukraine's power grid in 2015, cutting off electricity to 250,000 people. Almost exactly a year later, hackers caused another blackout.

It is now abundantly clear that cyber attacks present a danger to more than computer systems. Cybersecurity must now include the critical infrastructure—including utilities—that undergirds communities.

The War Comes to Germany

"The cyber warfare in Russia and Ukraine is one of the main interests in Germany because we saw a lot of similar hacking,"

Kozok said. "We see hacks against NATO and against German governmental and commercial systems. Our intelligence community has developed a wide-ranging response to this undeclared cyber war."

As hybrid warfare has evolved, experts like Kozok have developed a "hybrid intelligence" approach. In the past two years, Germany's armed forces refined their capabilities by using GIS to organize large amounts of intelligence data in the context of location.

"First, you have to bring together the raw data you have in different IT forms," Kozok said. "Then you have to combine it with the information from other sources. You need the cyber intelligence expert working in close cooperation with the geo-expert to find the best solution and visualization."

This map of the languages spoken in Ukraine shows the complexity of ethnic divisions.

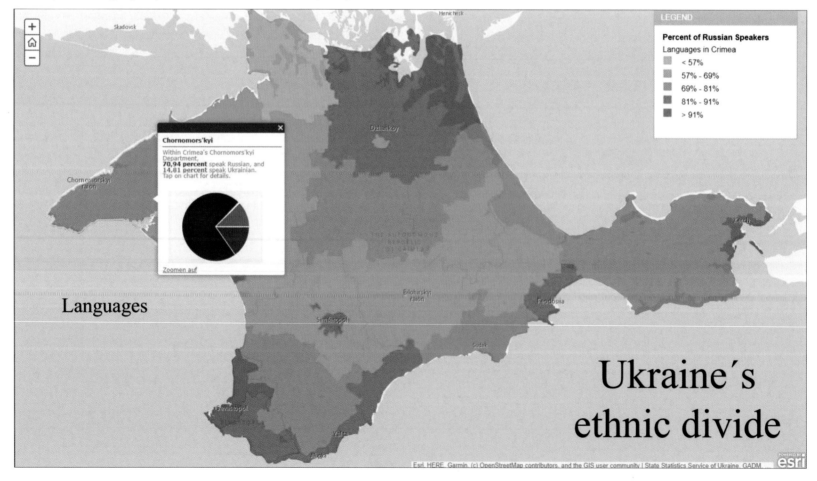

Location Intelligence Meets Cyber Intelligence

Data analysis takes many forms. It can be as basic as visualizing a terrorist attack against the backdrop of a city's underground infrastructure or as complex as plotting the spread of propaganda by using AI to sift through social media feeds.

The hybrid approach has proved especially useful in handling a barrage of cyber attacks against German interests in recent years. Many attacks appear to originate from Winnti, a group of hackers headquartered in China. By adding a location component to related data, officials can more easily assess the problem and act on it.

"If I show a general some source code, he won't understand what I'm doing," Kozok said. "But if I can show him on a map that a Winnti tool has been attacking certain parts of Europe, he might see that it's mostly in the European Union, or perhaps it's an attack on NATO, which means the military is probably involved. With another map layer, I can show him how many of

A ship carrying undersea cable is hard to conceal, which made the move by Russia to lay a direct line to Crimea apparent.

the attacks are against chemical companies."

Kozok has found that location intelligence provides a common parlance to understand cyber intelligence. "The generals understand that the world belongs on maps," he said. "It's how we show them the connections linking every analysis we have."

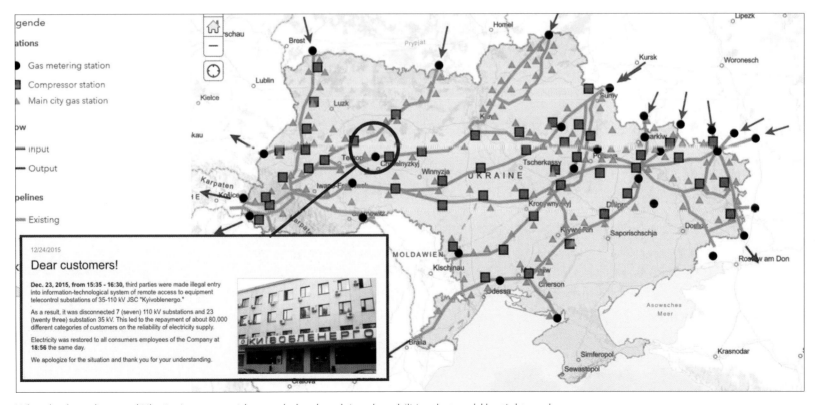

When hackers disrupted Ukraine's power grid, maps helped explain vulnerabilities that could be tightened.

Conservation and Environmental Protection

Sustainable development, as defined by the 1987 *Brundtland Report* published by the United Nations, is "development that meets the needs of the present without compromising the ability of future generations to meet their own needs." Many rightly point out that the stewardship ethos of Indigenous peoples precedes this sentiment, such as the Great Law of the Iroquois to "live and work for the benefit of the seventh generation into the future."

Regardless of the origin, to live and operate sustainably demands a conservation mind-set and the ability to take the long view. There are many lifestyle options to achieve individual goals to lessen consumption. For governments and organizations working to assess the sustainability of their complex systems, a toolset is required to measure change, map environmental health, consider impacts on all people, and pinpoint resources to achieve progress and surmount problems.

Sustainability practitioners have long used GIS technology to assess social and environmental impacts and devise plans for improvement. GIS contains the tools to analyze and model the health of ecosystems—from urban cityscapes to the remote wilderness. It also provides the means to understand the needs of vulnerable people and to build both social and infrastructure resilience.

Sustainability goals are increasingly becoming policy objectives of governments and municipalities. Pledges to meet or exceed conservation and emissions targets have taken many forms, marking an exciting shift to practices that will improve the stewardship of our planet.

Around the world, municipalities and countries have pledged to "go zero" on carbon, shift to renewable energy, achieve emission-free mobility and industry, and create circular economies by engineering the elimination of waste and pollution.

More than 70 countries have promised to conserve 30 percent of land and ocean by 2030 (30x30), on the road to Half-Earth goals. Foundational work has been done to build the data to inform conservation decisions, such as Esri's Green Infrastructure initiative to find areas in need of conservation and Nature Serve's Map of Biodiversity Importance to address imperiled species.

Managers of natural areas have developed GIS techniques to determine the suitability of habitat, preserve migration paths, and create wildlife corridors to ensure the greatest outcomes. In agriculture, GIS is being applied to maximize yield with the fewest inputs of fuel, fertilizer, and pesticides. Similar approaches are being applied to conservation, determining how not to just return lands to nature but to apply accelerants to rewilding areas no longer used by industry.

The UN Sustainable Development Goals seek to end poverty, protect the planet, and ensure prosperity for all as part of a new sustainable development agenda. This data-driven framework, with indicators measured and tracked using GIS, provides an approach for communities to emulate to improve the treatment of people and our natural world for a more sustainable future.

Professionals are applying GIS to all these goals, focusing on reducing emissions of greenhouse gases and setting aside areas to reverse the tragic loss of biodiversity.

We can all learn from the scientists, ecologists, and geospatial professionals who are mapping, measuring, and modeling to make areas more resilient. GIS is helping determine what works best, such as what trees to plant to sequester the greatest amount of carbon. Above all, GIS gives governments and institutions the means to monitor progress toward meeting conservation and sustainability goals and fine-tune policies to achieve ambitious milestones to preserve our environment.

Mapping America's Land and Sea: A Time for 'Precision Conservation'

By Dawn J. Wright, Esri Chief Scientist

The new Biden administration report, *Conserving and Restoring America the Beautiful*, is a welcome answer to the clarion call of several emerging global movements, chief among them Half-Earth, to stop the rapid loss of species and improve resiliency to climate change by setting aside more of the earth's surface for nature before it's too late. President Joseph Biden proposes preserving 30 percent of US land and ocean by 2030

(30×30) through "a 10-year locally led campaign to conserve and restore the lands and waters upon which we all depend, and that bind us together."

As a scientist, I'm excited the Biden administration recognizes the need for data and science to shape decisions and actions. As an oceanographer and geographer, I'm also excited because the most critical part of the mission—deciding

Farmers use sensors and maps to determine where interventions can improve crop health and yields.

where to conserve—is one for which we have the technology and approach to support. That specificity can be found in precision conservation, a relatively new methodology that is redefining how landscape and seascape conservation should be approached with smart maps ensuring that conservation projects are the right size, implemented at the right place, at the right time, and at the right scale.

As the 30×30 directive ushers in a new era of needing to work with nature rather than against it, we must reevaluate lands and seas with an eye on biodiversity, carbon emissions, and other environmental factors. We also must be mindful of issues of environmental justice as well as the interests of Indigenous populations, economic health, and local communities. To accomplish this, we can borrow from the fields and practices of precision agriculture and precision public health and make advances in precision conservation. These precision approaches are made possible by the range of data collection, management, and analysis supported by a modern GIS, which helps answer questions about the balance between natural systems and human-made systems.

Knowledge about people, place, and planet enabled by GIS will be crucial to ensure 30×30 actions are environmentally and socially intelligent.

During the coronavirus pandemic, we have seen how essential that science, public policy, and technology can be in tackling complex challenges to advance the greater good. We have learned that interactive maps and spatial analysis can provide greater awareness and context and that sharing data on a dynamic map allows diverse groups to act together with purpose.

As local and national leaders respond to Biden's *America the Beautiful* report, they will need precision conservation to meet the capabilities and intensity that this moment demands.

The location intelligence that can be applied through precision conservation can reveal pathways to smart conservation. For instance, 30×30 goals aim to set aside last-chance intact ecosystems before they are harmed. Working with a modern GIS, scientists can quantify available places, qualify an area's ability to enhance biodiversity or strengthen climate resilience, and provide critical context to help us understand the role of local and regional ecosystems in the bigger picture.

Meeting the Challenge

Just as the agriculture and public health sectors have been using maps and modeling through a GIS for better outcomes, the ambitious objectives of the 30×30 program can be guided by GIS technology, which helps us look through space and time to unlock new knowledge about nature and its ecosystems.

Advanced location intelligence from GIS is already empowering a broad range of conservation initiatives:

- The Pacific Ocean Accounting Portal spatially integrates public data about the protection, rehabilitation, restoration, and governance of the Pacific Ocean, capturing its real-time condition. It employs the System of National Accounts, from which gross domestic product and other economic measures are derived.

- In the Sahel region of Africa, GIS is used to pick the right species of tree to plant and to monitor the health of each sapling to give the Great Green Wall a chance of reversing desertification.

- In places like Palau, Micronesia, detailed dynamic maps show threatened reefs, addressing human activities that put pressure on these vital ecosystems.

In the US:

- The Map of Biodiversity Importance, or MOBI, was created by NatureServe, in collaboration with Esri and The Nature Conservancy to provide a comprehensive set of habitat models for more than 2,200 at-risk species, both flora and fauna, in the contiguous United States. It features AI "predictor layers" that anticipate species viability based on both development plans and environmental factors.

- The National Water Model, run by NOAA and driven by approximately 7,000 observational measurements, hourly precipitation forecasts, and landscape characteristics, estimates water flow on 2.7 million streams and reaches across the continental United States.

- Across the largest US estuary, the Chesapeake Bay, GIS provides a way to communicate and share successful strategies among jurisdictions and states that contribute to water quality.

Today, entirely new kinds of maps and data visualizations are made possible by the instrumentation of natural and human-made systems and the integration of many types of data. This enables radically enhanced contextual awareness, which lets us understand and optimize issues and entire complex systems.

As we've seen with the pandemic and as we're discovering with issues such as climate change and sustainability, this greater contextual awareness is critical to our decision-making and problem solving. Importantly, GIS is the technology of contextual analysis. GIS not only integrates and analyzes diverse datasets, but it also brings the physical world, including people and place, into the core of that analysis, enabling us to see the interconnectedness of our many systems.

Changing Our Mind-set

The Biden administration's 30×30 effort presents an unprecedented opportunity to unlock and integrate datasets that describe the natural world and that will enable a decision support system for preserving our country's rich biodiversity.

The Great Green Wall in the Sahel region of Africa is designed to arrest increasing desertification.

The reefs of Palau, Micronesia, are considered the country's greatest asset.

By applying nature-based conservation strategies, 30×30 aims to slow and reverse the environmental degradation that has led to species decline and extinction. This idea of designing with nature has long been a guiding focus of GIS, a technology that works by layering large varieties of data on an interactive map for deeper understanding.

GIS can capture a wealth of expert knowledge in a single place: a hydrologist's understanding of stream flow, a geologist's understanding of earth structure and process, a botanist's knowledge of plant physiology and classification, an oceanographer's understanding of fish migrations and hurricanes, a policy maker's understanding of implications, and the domain expertise of many more specialists and generalists.

The technology can also integrate with advancements in earth observation and AI, adding more regular and precise measurements and new monitoring capabilities. Drones, for example, provide flexibility, new perspectives, and greater sensing capacity. And anyone can contribute—with GIS workflows, mobile phones can be used to capture data and share knowledge about a place.

Sharing maps, models, and GIS-based analysis within a 30×30 decision support system will empower stakeholders to accelerate action and ensure sustainable impacts.

In short, we have the mapping and analysis tools we need—and are ready to support our leaders as they take on this grand challenge.

Applying Science and Mapping to Safeguard Species: Half-Earth Project

Maybe you've dreamed of catching a glimpse of the ivory-billed woodpecker in a southern swamp. Now the best you can do is watch a video, because in October 2021, scientists at the US Fish and Wildlife Service declared this and 22 more species extinct. The cause of habitat degradation and destruction has raised fresh cries to set aside more land and ocean for nature.

Decades ago, Robert MacArthur and E. O. Wilson pioneered a theory to assess how larger conserved land areas support the survival of more species. Their original theory, developed in the 1960s, was born of research on island biogeography with the observation that larger islands contain more species, but it has since been applied more broadly to species preservation.

"The theory addresses biogeographical dynamics," said Walter Jetz, professor and director of the Yale Center for Biodiversity and Global Change. "Larger habitat fragments and those closest to other undisturbed forest habitats hold more species, and as humans encroach and create smaller patches that are less connected, you rapidly lose species diversity."

Wilson articulated the theory in his 2016 book *Half-Earth, Our Planet's Fight for Life* where he proposed, "Only by committing half of the planet's surface to nature can we hope to save the immensity of life-forms that compose it." In addition to reducing the human footprint, Wilson also called for a deeper understanding of ourselves and the rest of life on our planet.

habitats and their viability to support a variety of species. The goal is to support the right decisions and investments to curtail the current mass extinction.

"When you lose species, you lose important functions and services," Jetz adds. "At some point, the key functioning—the intactness and integrity of the ecosystem—isn't there anymore because too many species have been lost."

Wilson joined special guest Sir David Attenborough, along with Sir Tim Smit of the Eden Project at The Royal Geographical Society, in London in October 2021 for a livestreamed global event for the fifth annual Half-Earth Day to take stock of the world's progress.

As Wilson puts it in *Half-Earth*, "People understand and prefer goals. They need a victory, not just news that progress is being made. It is human nature to yearn for finality, something achieved by which their anxieties and fears are put to rest."

The E. O. Wilson Biodiversity Foundation's Half-Earth Project, with Jetz as its scientific director, aims to tackle both of Wilson's objectives of quantification and conservation—saving half the earth while safeguarding biodiversity. This global biodiversity conservation project has set out to quantify habitats and the geographic distribution of species—putting both on the shared open Half-Earth Project Map. The effort also looks at how land cover change and climate change affect

The nocturnal oilbird makes its nest in caves and feeds on fruit, mainly the nuts of oil palms.

Birds Provide Proof of Precariousness

There is perhaps nothing more troubling than losing a species forever, which is what has happened—again—with last fall's news from US wildlife officials. When we lose a species, there is often a disturbing ripple effect. Jetz, for example, points to the endangered oilbird in South America, the only nocturnal flying fruit-eating bird in the world. It serves as a major seed disperser in the forest, and if it were to be lost forever, we would likely see changes in forest structure.

Birds became Jetz's obsession as a teenager, and as an undergraduate student, he started using GIS technology to map and analyze their distribution. He went on to create the first global map of bird species during his graduate work.

"I've always been fascinated by maps, and then I got really fascinated by species and biodiversity and how environmental gradients change species composition," Jetz said. "There's a lot of information about birds, and it's a reasonably small species group with roughly 10,000 species globally. It was the first species where it was possible to grapple with issues of global biodiversity and the implications for conservation science."

Since Jetz's early work in the 1990s, global maps of species distribution have expanded to mammals, amphibians,

The Half-Earth Project Map was developed by the Map of Life Project at Yale University's Center for Biodiversity and Global Change with support from the E.O. Wilson Biodiversity Foundation, the science and technology company Vizzuality, and Esri.

GO TO EXPLORE DATA

HALF

Priority places for biodiversity

The highlighted areas show places of particular importance to life on Earth. Understand these landscapes and how they need our care.

ALL MAPS >

all species　　reptiles　　mammals　　amphibians　　birds

ABOUT THE HALF-EARTH

plants, invertebrates, and marine life. Putting this work into a shared map for collaborative science is now Jetz's passion, to avoid duplication of effort and to pool efforts to arrive at a scientifically rigorous yet actionable knowledge base.

"Alexander von Humboldt was the first to draw some of these maps and gradients 250 years ago," Jetz said. "Now it's possible to take that into the quantitative scientific realm and to think not just about single species but to make the link to patterns of communities and the conservation relevance of places."

Ecoinformatics to Measure Change

Increasing volumes of data are helping underpin spatial decision-making to protect species at the local, regional, national, and global scale.

"We have the tools, and increasingly the science, to address this issue," Jetz said. "We can take species into account, and we can include them in our planning and decision-making."

The spatial dimension unlocks patterns of a species richness and rarity, including how their traits function across geography, how they relate to the history of an area, and where they are native and live in large numbers.

Determining the optimal area to preserve entails looking at hundreds and thousands of species in a region to identify priority places and think efficiently about safeguarding a maximum number of species in the least amount of area.

The project has created maps that weigh the importance of places, including metrics on human encroachment, existing conservation protections, and details about the species.

"It's not simply the hot spots of richness that we need to conserve but the right optimally designed network," Jetz said. "In the Half-Earth Project science, we use the species distribution information to give each location a globally informed, quantitative priority score. It's a dynamic map that changes as our knowledge progresses and our conservation progresses."

Taking Action to Make Progress

The simplicity of the Half-Earth concept has helped it gain momentum, with core innovation in how to best care for our planet being guided by rigorous science. Both sides—conservation and quantification—have increasing urgency if we're going to preserve biodiversity for future generations.

"The guiding mission of the Half-Earth Project is to leave no species knowingly behind," Jetz said. "Extinctions will continue, but we shouldn't be losing species unknowingly. We have a moral obligation, at a minimum, to be aware when we drive a species to extinction and to know how to safeguard every species."

E. O. Wilson pioneered a key theory that relates the carrying capacity of land and ocean to preserve biodiversity.

Maps Cut through the Fog in Peru to Help Preserve Unique Ecosystems

High in the foggy Peruvian hills, for just a few weeks in June, bright-yellow flowers resembling tiny phonograph horns light up the hillsides. Known as the Flower of Lima, its naturally short, seasonal life and the fog oasis it calls home are increasingly under threat.

Illegal land-grabbers have seized on an outsized demand for affordable housing as more Peruvians are pushed farther from Lima's established settlements. These land traffickers, having staked and sold fraudulent claims to the soil that the Amancaes (pronounced: ah-mawn-KIE-us) flower calls home, has led those desperate for housing to build structures on land that not only needs protecting but can be vulnerable to earthquakes and poses health concerns because of pervasive humidity. The unmanaged development, including unsanctioned quarries where miners extract construction materials, has left behind damaged land unfit for plants.

The unique flora and fauna that make up the fragile ecosystems of this region's lomas, or fog oases, have long been at risk of disappearing. In late 2016, the Lomas EbA project, an initiative directed by the National Service for Natural Protected Areas (SERNANP) of Peru and implemented by the United Nations Development Programme (UNDP) with support from the Global Environmental Fund (GEF), was born to preserve 19 of the estimated 100 lomas that only occur along the desert coasts of Peru and northern Chile.

Settlements creep ever closer to the fragile ecosystem. (Photo courtesy of UNDP Peru)

"We can't lose the only place where this amazing flower grows," said Adriana Kato, a communications specialist with the UNDP group working to preserve the lomas.

Map-Based Activism

The disappearing fog oases in this high-elevation desert are at the crux of a demand for housing with about one-third of Peru's population, nearly 11 million people, living in Lima. Although poverty levels in Peru have dropped from nearly 59 percent to about 22 percent between 2004 and 2015, much of the population remains on the cusp of falling back into poverty. With a lack of housing strategy in Lima, unorganized development has encroached into the fog oases.

"It's not just an environmental problem," Kato said. As a recent UNDP report noted, the effort is aimed at addressing an "unprecedented complex combination of problems."

With so many living in Lima, one might assume it would be easy to find environmental defenders. But the hills faced a unique challenge: few people have witnessed the land in its full glory because it's covered in a blanket of mist from May to October. By the time the fog lifts, the land turns back to dry desert. Kato, who grew up in Lima, was among those who didn't know about the nearby, hidden beauty until she joined the UNDP.

"Lima is surrounded by these fog oases, these green hills, but not many people know about them," Kato said. "We have all this, but we don't see it."

The UNDP's Maria Miyasiro, a GIS and remote sensing specialist, started raising awareness by building a detailed geoportal to share local apps and maps, including one named GeoLomas, to allow people to explore the areas that had gone unseen. Local environmental leaders, often self-taught experts in botany and laws relating to the lands, have gathered data for GeoLomas using the ArcGIS Survey123 app on phones and tablets to input what they observe as part of a pilot project.

"I decided to build a web app so our stakeholders and our environmental leaders could do their own analyses and make their own maps," Miyasiro said. "We started with just one application, the GeoLomas web map viewer, and then we elaborated with more applications created together with the stakeholders, organizations, and municipality workers, and using ArcGIS StoryMaps to publicize their work."

Volunteers have been adding layers to the GeoLomas map, including archaeological and cultural features, potential tourism locations, and land rights, to know when an area was being illegally encroached on. The group worked closely with park rangers to record flora and fauna species they encountered.

The online map was described in a recent midterm review of the project as the "biggest advance" toward tracking conservation actions and improving transparency.

Sustainable Development around Lima

The UNDP team also set out to tackle several of the UN's 17 Sustainable Development Goals: improving the environment for those living on the edge of Lima where there has been less access to green space, creating a sustainable and resilient city, conserving the land and biodiversity of hill ecosystems, and reducing poverty by creating jobs in ecotourism and agriculture.

"It's really necessary that people understand that the lomas are not the place to put their houses," Kato said. "It's really dangerous because of earthquake risk and for their health because of high humidity. It's better to conserve them, and make a really nice and safe path, so people will visit them and use it for ecotourism, encouraging community members who may have otherwise lived in the oases to instead preserve the lands and launch enterprises for paying visitors."

In an audit of the lomas project's progress, a report noted that despite the complexities faced, "the balance sheet is positive in terms of will to participate." In Lima, about 10 different local organizations now care for their nearby fog oases, in some cases starting nurseries to nurture new plantings and using a fog-catching mechanism to collect water for irrigation.

Before the creation of the online map, many didn't know the physical boundaries of the land they were protecting. Now, on a smartphone using the ArcGIS® Explorer app or directly in the browser using the GeoLomas web app, they can see whether a house has been illegally erected inside the border or if a new road and plots for sale belong to land-grabbers, requiring them to call the police.

During the pandemic, outreach and training for contributions to the GeoLomas map shifted even further online, and leaders have leaned on more technology. Drone footage of the

Noe Neira, a leader of the Lomas de Paraiso citizens' conservation group, uses a map to show what they're defending. (Photo courtesy of UNDP Peru)

Volunteers use maps to raise public awareness. (Photo courtesy of Haz tu mundo verde, the volunteer group pictured)

team has also partnered with the Geological Institute of Peru to preserve unique geologic formations.

By collecting key data and mapping it with GIS, advocates were able to communicate the location of each fog oasis, identify local flowers and animal species, site nearby archeological locations, and offer details about pedestrian access.

"It's a really, really valuable tool for us," Kato said, for as she and Miyasiro wrote in the interactive map they authored, "You don't take care of what you don't love, and you don't love what you don't know."

land has become a primary feature on the maps, helping raise awareness with immersive video that allows people to fly over the scenic hills.

In a little more than four years, the work of organizers and the volunteer force of residents helped map and catalog information about this area. Details gathered provided the evidence needed for formal preservation of at least five of the oases, which are now part of a newly designated regional conservation area called Sistema de Lomas de Lima.

Now, the UNDP is devising a strategy for handing control of conservation efforts back to Lima's government, including bringing housing, cultural affairs, and law enforcement officials to the table.

The team encourages private conservation developments and works with the Ministry of Culture to protect other fog oases that include archaeological features at risk of being lost. The

Nataly Julca, a geographer and enthusiastic volunteer, uses the GeoLomas map viewer on her phone to check the border of the ecosystem. (Photo courtesy of UNDP Peru)

Drones Key to Mapping Massive Barrel Dump Site Off Southern California Coast

The research crew that had hastily assembled off the coast of Santa Catalina Island in Southern California, some 3,000 feet above the ocean floor, could hardly believe what they were seeing. They knew they would find dumped barrels resting on the seabed—they just didn't realize how many. Years earlier, scientists had detected dichlorodiphenyltrichloroethane (DDT) pesticides seeping into underwater sediment and found traces in the fat of fish. A targeted dive even provided alarming video evidence of barrels degrading, ringed by fluorescent traces of leaked materials.

Now, the technology was available to do a comprehensive survey to determine just how sprawling the toxic graveyard might be, using a GIS to lay it out for anyone to see on a map.

For two weeks in March 2021, researchers from the Scripps Institution of Oceanography at the University of California (UC) San Diego worked day and night mapping 46 sq. mi.

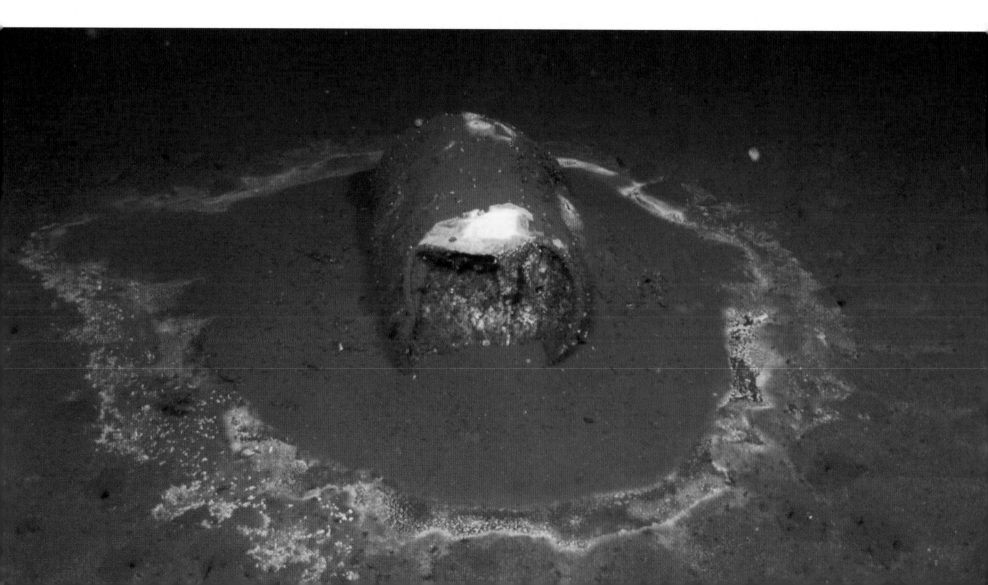

of the ocean floor between the island and the coast of Los Angeles with the help of two underwater autonomous vehicles equipped with sidescan sonar that could skim the bottom of the sea, emitting sound waves and using acoustics to detect any sign of barrel-shaped objects on the seafloor.

What they found: more than 27,000 of the metal cylinders resting on the seafloor. That figure included many arrayed in a line as far as 11 mi. between San Pedro Harbor and the island, indicating that the barrels may have been methodically dropped off a barge or moving boat.

New Technology Uncovers Known Problem

"In hindsight, we shouldn't have been surprised," said Eric Terrill, who led the mission as chief scientist and directs the Marine Physical Laboratory (MPL) at Scripps. "But it's one thing to read historical reports that there may be tens, if not hundreds of thousands, of barrels on the seabed. It's another one to start counting those in acoustic sidescan data," he said.

That large number represents only what they were able to map in a short window of time.

Although 46 sq. mi. may sound like a significant area to cover, and it is, "there's a lot more area of the San Pedro Basin to go," said Heidi Batchelor, a GIS analyst at Scripps who participated in data processing on the expedition. In the end, the crew ran out of time that was funded and allocated to them for use of the ship but "not out of debris."

Montrose Chemical Corporation was the United States' largest manufacturer of DDT pesticide, and for years it was suspected of dumping barrels in addition to the chemicals it sent into the ocean via sewer pipes. The popular assumption, at the time, was that the pesticide would be diluted and dissipate in the water. It didn't.

In a sprawling *Los Angeles Times* investigation last year, the newspaper discovered the existence of shipping logs showing the company had dumped more than 2,000 barrels a month into the Santa Monica Basin in the years following World War II, estimating there could be as many as half a million on the seafloor.

In less than five months of preparation, researchers boarded the *Sally Ride*, the newest addition to Scripps' fleet of research vessels, to see what was below, at the urging of federal lawmakers and the National Oceanic and Atmospheric Administration (NOAA). NOAA's Office of Marine and Aviation Operations and the National Oceanographic Partnership Program collaborated on the project.

"Normally, these vessels are distributed across the world doing global oceanographic science. So we were fortunate to actually have the *Sally Ride* available to us in Southern California," Terrill said.

Scripps researchers aboard the research vessel *Sally Ride* deploy the Bluefin autonomous underwater vehicle to survey the seafloor for discarded barrels. The expedition was a 24-hour operation in March 2021. (Photo Credit: Scripps Institution of Oceanography at UC San Diego)

Even as DDT's popularity for combating mosquitoes and agricultural pests remained strong, there were stark reminders from scientists that no one really knew the prospective long-lasting effects of the chemical on the natural world. Biologist Rachel Carson's *Silent Spring* warned of a future without songbirds and other species as the chemical wreaked havoc on a world not properly protected. The book helped usher in the modern environmental movement in the 1960s. By 1972, the US had banned DDT's use. However, there was still demand from other parts of the world, and Montrose kept operating until 1982.

Years later, first in 2011 and again in 2013, UC Santa Barbara Professor David Valentine discovered concentrated levels of DDT in ocean sediments between Los Angeles and Santa Catalina Island, and cameras captured images of at least 60 leaking barrels at the site where Scripps researchers focused their search. They set sail in mid-March to see how many barrels they could find.

"Ten years ago, it would have been impossible to do the type of work we did on this expedition because the technology just wasn't there," Terrill said.

Scripps researchers inside the operations center aboard the research vessel *Sally Ride* during a seafloor survey in March 2021. (Photo credit: Scripps Institution of Oceanography at UC San Diego)

No Gray Area

Despite amassing more than 100 gigabytes of sonar data, the dataset itself wasn't so big that it couldn't be managed. The challenge was that there were a tremendous number of barrellike objects in a relatively small area, Batchelor said. Researchers didn't want to overlook anything or risk counting the same barrel twice. They also wanted to ensure that they counted and mapped just what were barrels and not other objects. They sought help from Scripps colleague Dr. Sophia Merrifield, an associate researcher who specializes in machine learning, to accurately characterize objects in imagery. And with her help, their results greatly reduced anomalies.

"We were able to immediately rule out any kind of electrical noise, computer glitch, or issue with sensor settings in the data," Batchelor said of their confidence in the results. "There's no gray area there. There are objects on the bottom." And those objects are all consistent with the size and shape of barrels such as those used for dumping DDT. Exactly what the barrels contain, whether DDT or something else, will require separate research.

Although the search for barrels turned out to be far less of a hunt because there were simply so many, mapping them still proved challenging because of the need for accurate results.

"Resolution often matches to the scale of map you might be using. If you're looking at a map of the US, you aren't looking at individual street signs. But in this case, we needed to actually get that granularity to be able to map out objects that were just a few feet in size," Terrill said.

That meant as much as 2 to 3 cm metric abbreviations used in at least one earlier article, so changing here resolution in portions of their survey area, a uniquely high level of resolution considering the large area mapped, to ensure the accuracy needed for the work they were doing, Batchelor said.

One Picture Tells the Story

There was a heightened sense among the crew that the expedition would require solid analytic tools to defend the provenance and quality of the data.

"One thing we wanted to be very careful about is not alarming the public. As scientists, it's our responsibility to have defensible data provided to the public and the decision-makers," Terrill said.

"Really understanding the spatial distribution of the dumpsite was very important to us here," Terrill said, as was communicating the findings. That included a GIS heat map of barrel locations providing an accessible depiction of where the highest density of debris existed.

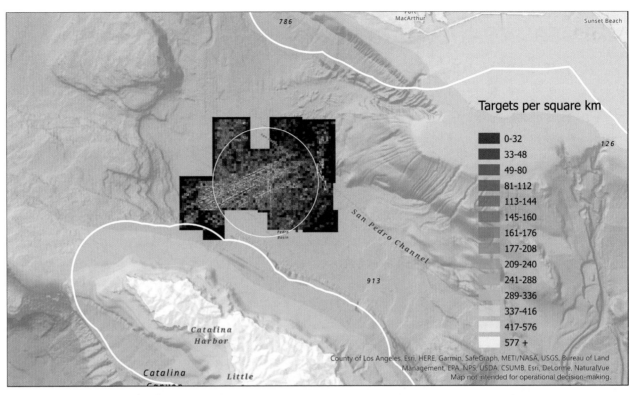

A heat map showing concentrations of targets detected in the San Pedro Basin. The presence of several distinct track-line patterns in the surveyed area suggests that the dumping was repeatedly done from an underway platform such as a moving ship or barge. (Image Credit: Scripps Institution of Oceanography at UC San Diego)

The first step was a typical spatial analysis, using the "nearest neighbor" test to see if the barrels were clustered together or not.

"They were very clustered," Batchelor said. "They're in linear batch areas and big concentrations, but we needed to prove it, so we ran the data through some spatial statistics."

When you have more than 25,000 potential objects to plot, though, it can be difficult to see, so Scripps analysts made a grid.

"Then we were able to do a count per fishnet grid cell. And then once you have that, you have free rein to visualize it however you like," Batchelor said.

"In the end, you get one picture."

That picture, in the form of a map the team created with GIS, has helped the public understand the magnitude of the problem.

As for the problem itself, it won't be solved overnight. Scientists are still in the early stages of understanding what lies in front of them and what will need to be done, including how or if any of the toxic contents on the seafloor will be exhumed and determining the contents of the barrels. "We're starting to get a strong cadre of scientific engagement to address the broader holistic issues at the study site," Terrill said.

"All along, we went into this trip knowing that the data we collected would inform a broader strategy of response to the area. And so, what we have in hand now is the ability to know where to go next."

Mapping to Protect Natural Heritage and Biodiversity in South Carolina

To understand the complex variables and changes in natural cycles that impact threatened species, conservationists are using sophisticated mapping techniques and GIS technology.

Consider the gopher tortoise and its unique relationship with forest fires.

The gopher tortoise is resilient to the occasional forest fire because the long tunnels it digs and inhabits 10 feet underground are fortified enough to withstand the smoke, flames, and burning debris.

Maintaining this subterranean lifestyle requires specific habitat conditions. The sandy soils of the pine savannas found throughout the southeastern United States are particularly welcoming.

Over time, as the pine trees grow, the forest density threatens the vegetation the tortoises eat. Before major human contact, naturally occurring forest fires—often caused by lightning strikes—would help preserve this understory vegetation and protect the tortoise's diet.

The vegetation a gopher tortoise eats requires open land, which naturally occurring forest fires help maintain.

The modern fragmentation of the forests—through the introduction of roads, homes, croplands, and cities—has disrupted this cycle. Without fires, the forested areas that remain have become overgrown.

The decline of fire-adapted forest ecosystems is not the only factor that threatens the gopher tortoise, but it underlines the complexity of the fight to protect the endangered creature. The South Carolina Natural Heritage Program is arguably its most powerful ally.

Protecting South Carolina Heritage

The Natural Heritage Program—and its Cultural Heritage counterpart, dedicated to preserving land with historical and cultural relevance—together form the South Carolina Heritage Trust Program, a section within the state's Department of Natural Resources (SCDNR).

The work of the Heritage Trust began in the mid-1970s. Shortly thereafter, when the project was transferred to SCDNR, the state agency became the first state in the country dedicated to protecting land with abundant natural or cultural significance.

From the start, the work relied on one of the earliest desktop GIS software programs. As GIS software grew more sophisticated, Heritage Trust adopted an ArcGIS Enterprise approach. The technology became central to many of the program's objectives, including mapping habitat loss and picking areas the state could purchase for permanently protected heritage preserves.

With GIS, conservationists can analyze how and where to develop land in ways that protect biodiversity.

Shortening the "Long, Drawn-Out Process"

The Natural Heritage Program recently expanded its GIS to facilitate better communication with stakeholders, including private developers, scientists, and the public. Communication and workflows have always been challenges, especially regarding due diligence for developers proposing new projects. Any development that receives federal government financing or permitting must review the presence of threatened or endangered species within the proposed project footprint.

Until recently, developers would submit requests with varying degrees of specificity. Some contained only geographic coordinates, others included maps, and some were just general descriptions of the area. The office would check requests against its database of protected flora and fauna, struggling to get through around 200 requests per year.

"It was a long, drawn-out process, and it wasn't even our full-time job," said Joe Lemeris, the program's GIS and data manager.

The inefficiency extended to the workflow used by scientists who gathered data from the field. They recorded their observations onto Excel spreadsheets or paper forms. At a later date, this information would become part of the program's database. The office also gathered data from other civic and private partners, including the US Fish and Wildlife Service, which could arrive in various formats, adding more bottlenecks to the process.

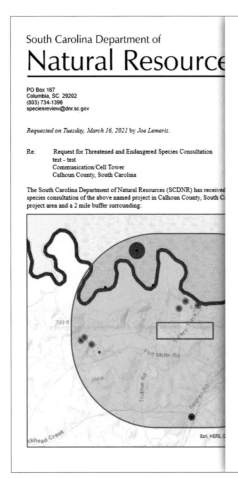

A sample SCDNR report details the presence of endangered species and shares best practices to minimize impacts.

"The problem with the way it was done before is that the consultants could say they'd done due diligence, and that we'd shown them everything we had on a species, but then more data could pop up after we'd shared what we had," Lemeris said.

Moving Beyond Static Data for Solar Power

A few years ago, a proposed solar project northwest of Beaufort, South Carolina, put a spotlight on the outdated system. The developers consulted information previously obtained from the Natural Heritage Program database, unaware that data gathered more recently showed that current project plans threatened the gopher tortoise's habitat.

The problem, as Lemeris saw it, was that everyone was dealing with "static data"—information frozen in time as the world moved forward. What the program needed was a dynamic system that would reflect the most recent findings.

The solution was to take advantage of recent developments in GIS that prioritize the storage and flow of geographic information as well as its visualization. Data is now added and accessed through the same GIS-enabled portal. "It's essentially just a web app, and we added a custom-designed reporting tool," Lemeris said.

Scientists conducting field surveys input their data into ArcGIS Survey123 forms connected to the database.

"They have access to the data being collected in a way that goes beyond just writing it down in field notes and then having to transcribe it into a spreadsheet," said Tanner Arrington, GIS manager for SCDNR. "They gain a locational context that they didn't have when the data was just tabular. With location attached to it, they can see patterns that were unknown before."

Protecting Species from Foot Traffic

The GIS data can be presented with varying degrees of specificity, depending on who wants to see it. For certain species, the Natural Heritage Program prefers not to share its survey data in granular detail, opting instead for a more generalized view.

The black rail bird merits this opaque approach. This bird's tiny stature restricts it to wetland areas with less than an inch of water. Unlike most birds, they spend more time running through marshes than flying through the air.

For many years, no black rails were sighted in South Carolina, and the bird remains on the threatened list under the Endangered Species Act. Recently, birdwatchers have logged black rail sightings in the state. Although this is good news, birder enthusiasm can present a problem.

If the exact location of appearances by the black rail are revealed, the influx of birders may damage those spots. On the Natural Heritage Program's maps, only those with special permissions—scientists, mostly—can study exactly where rails have appeared. Others, including developers, will see general areas marked off as the bird's habitat.

Starting a "Ripple Effect" with Enterprise GIS

The success of the South Carolina Natural Heritage Program's portal has already paid dividends in increased efficiency. Lemeris estimates the tool has quadrupled the number of requests the office can process. The ease of use also encourages more developers to submit requests for data.

"I think there were a lot of people who weren't submitting projects to us because it was slow, and every day is more money," Lemeris said, "and also because it wasn't made very easy for them."

Other offices within SCDNR, as well as other state natural heritage programs, have also taken notice. "There's been a ripple effect," Arrington said. "The original understanding of the technology was through desktop GIS. But now that more people have seen what enterprise GIS is capable of, it's opened new opportunities within the agency. They're saying, 'Well, if it works for the Heritage Program, why can't we give it a try?' They've seen the benefits firsthand, and that's very powerful."

A male eastern black rail offers an insect to a female during courtship. (Photo credit: Christy Hand/SCDNR)

Health

It's remarkable to think that a map was one of the first indications that COVID-19 would become a global pandemic. The Johns Hopkins COVID-19 Dashboard gave the world an awareness of the virus, and we haven't taken our eyes off of it.

Quickly, communities around the world followed this pattern and started to track their own cases and hospital capacity using similar maps and dashboards. Maps were central to understanding supplies of personal protective equipment and how to shuffle masks and gowns to where they were needed most. The tools were in play to find and select optimal places for additional hospital beds, adding bed capacity in anticipation of a surge of spiking cases.

Public health agencies greatly improved their ability to find the vulnerable and deliver services more equitably. Epidemiologists were among the early adopters, and they used a host of ArcGIS tools and apps to identify those most in need. In rural Iowa, an epidemiologist learned of cadres of tight-knit immigrant communities working in meat packing plants, speaking more than 50 languages among them. Soon those communities were mapped, and volunteers were recruited that spoke those languages, enabling everyone to receive key information about the threats to their health and how to safeguard themselves.

GIS became especially important to plan and execute the massive vaccine distribution effort in the US and around the world. Web maps and apps overlaid information about vaccine delivery sites and population demographics. Geographic accessibility was calculated across multiple modes of travel, and again, populations with relevant vulnerabilities were identified and prioritized in resource allocation decisions. Dashboards saw a resurgence as people flocked to see the status of vaccine distribution compared with case counts, and to learn more about when they could get a shot and where.

The complexity of COVID-19 response has continued to increase along with the pandemic. We saw the rise of the Delta variant, a more transmissible form of the virus, reverse some of the progress that was made. The addition of booster shots impacts the counts of vaccines delivered and threatened to confuse the numbers of total Americans vaccinated. In an environment with too much distrust, many residents have taken comfort in the consistent reliability of maps to provide up-to-date situational awareness.

Maps have helped tell the story, including community engagement apps in which anyone could proclaim when, where and why they got the vaccine. Seeing use of the vaccine spread geographically and across age groups has provided a sign of hope, even amid setbacks.

As people began to take part in social and cultural activities again, GIS tools assisted in planning and managing physical distancing. As cases began to increase, GIS users analyzed human mobility data, looking for locations of concern.

An indoor view has helped companies reopen and respond should an employee become infected. A real-time map of indoor spaces allows a business to monitor sanitization operations and react to health and safety concerns.

The COVID-19 pandemic, and the utility of GIS maps and dashboards, has public health professionals around the world thinking broadly about spatially focused disease surveillance methods such as emerging hot spot analysis and wastewater testing. Global, regional, and local disease trend dashboards and maps will not only help with COVID-19, but they'll improve our overall understanding of diseases and how they are spread.

Beyond infectious diseases, GIS has long supported the needs of health professionals. In many places, such as San Bernardino County, California, health departments are extending what they've learned from COVID-19 about the people and communities they serve to deliver better service for all public health programs.

Finally, the pandemic has taught us some big-picture lessons—that everything humans do is interconnected and related to health, and that a health-in-all-policies approach is warranted and necessary.

COVID-19: An Iowa Epidemiologist, Maps, and a True Passion for Public Health

When the COVID-19 pandemic hit, epidemiologist Amy Hockett knew early on that what she was witnessing would trigger unparalleled needs for her county in Iowa.

Linn County, home to Cedar Rapids, is the state's secondmost populous county, with more than 225,000 residents. By early March 2020, the mounting number of hospitalized victims in the county because of the pandemic had already stretched medical resources.

More alarming, the region's first responders were falling ill, depleting the main core of those fighting COVID-19 at a time when they were needed most.

As the pandemic shut down schools and businesses, many people needed help securing food and shelter. The problems were compounded as local meat-packing plants slowed or closed, and employees—including many recent immigrants—found themselves without a livelihood and with

limited knowledge about where to turn for help. Anxious parents, teachers, and community leaders sought clear, concise, and authoritative information.

That's when Hockett decided she had to act. Hockett realized the crisis could devastate her community. In her eyes, what the community needed was someone to successfully assemble all the relevant data to help leaders coordinate their responses and provide resources where they were needed most.

Although she did not have extensive training in GIS technology or the location intelligence derived from geographic analysis, Hockett understood that it would be an extremely useful tool for compiling, analyzing, and visualizing data and trends and communicating that insight in ways that everyone could understand. She also believed that, considering the depth of the crisis, she had to expand her sense of what an epidemiologist could do in a moment like this.

"I never thought I'd be part of the emergency response," Hockett said. "In my everyday work, I'm at my desk, and I get to look at data. But that human aspect is always really important to me—being able to help anybody in any way that I can."

To get community members the data they needed, Hockett decided to create a COVID-19 dashboard. At first, the goal was simply to deliver basic information to members of the public, so they could visualize the spread and assess the risks. "How many cases do we have today? Who is being impacted? That was our very early goal," Hockett said.

Her work was inspired by the work of people from Johns Hopkins University, who had then started compiling national and international statistics on the spread, incidence, hospitalizations, and deaths related to COVID-19.

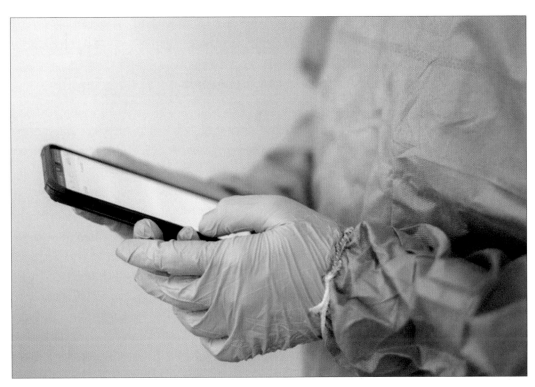

A health-care worker cloaked in personal protective equipment uses a smartphone to get updates on COVID-19 cases.

Making Mission-Critical Information Accessible

Like the Johns Hopkins effort, the Linn County dashboard was to be publicly available, easy to understand, and with data streaming in near real time.

"Our real goal was to give information to our public and partners," Hockett said. "So, one of the things we wanted was to be able to paint the local picture, the state picture, the national picture, and the international picture and make it accessible."

No one had ever attempted such an effort on a county level in the state of Iowa before.

The publicly accessible data collected eventually ran the gamut from simple spatial statistics to analysis of vulnerable people and rates of infection across space and time. Hockett and her team used GIS to layer data about food resources, ethnic groups and languages spoken across the county, homes where electrical outages could stop lifesaving medical machinery, and more.

That work has added immensely to the county's resiliency—and not just for COVID-19. In August, the data-rich maps also supported local responders in the aftermath of a derecho storm with hurricane-force winds that caused widespread destruction.

"All of our trees went down here, and it was terrible," Hockett said. "We have a great community here in Iowa, everybody wants to help each other, and that meant a lot of unmasking and a lot of close contact as everybody was cutting trees, removing debris, even helping people get out of their homes who were trapped by falling [tree] limbs."

Mapping Urgent Needs with Location Intelligence

Amy Hockett

Hockett's work to create user-friendly dashboards—which could keep up-to-the-minute tallies on vital indicators, engender public trust, and enable the public and first responders to grasp the reality of what they were facing with COVID-19—earned her national honors from the Washington, DC-based American Public Health Association with the 2020 Innovation in Public Health Award in her name.

Hockett recognizes the contributions of many public officials, agency leaders, and location intelligence specialists who provided data and helped her create the charts and dashboards that provided critical information. And she praises the quick adoption of digital and interactive smart maps to guide actions, including rallying resources to areas with the most pressing needs.

Hockett also can't help but call attention to the cooperation and partnership evident throughout the county. "We've cultivated a lot of different relationships over time," she said. "But I think this made it more clear to our community partners that we needed to work in a more efficient manner. We really needed to be able to share."

In giving her the award, the judges said Hockett's work helped community partners collaborate to ensure basic needs were met. And, they noted, "without the development of these innovative technologies and techniques, the response by Linn County Public Health would not have been as robust as it was."

Creating a hub for COVID-19 metrics with analyses down to zip code level required long days of work and cooperation with many departments, agencies, hospitals, and long-term care facilities. It also required a strong commitment to continually refine and simplify data presentation after receiving questions and feedback from the public. All of this was carried out under stressful conditions as new challenges were constantly presenting themselves, Hockett said. A shared understanding of the situation, delivered by the dashboard, allowed many in the community to feel they were on the same team and striving toward the same goals.

"We actually had a lot of different audiences that we had to consider," Hockett said. "We needed to be able to see emerging information, our hospital capacity, the trajectory of the disease, who the population was. All of that needed to be immediately accessible.

"In the beginning, we wanted to just make the case for information," she said. "But we had early difficulty with an inconsistency in our state-level data and our local data. So, our public really didn't have a lot of trust. That was where we wanted to step in and say, 'No, we want you to have the best information possible and in a way that you really can understand.'"

Digital Tools Help Visualize Community Needs

Her two main tools for collecting all the information and displaying it on smart maps were ArcGIS Survey123, an app that allows agencies to input data daily, and ArcGIS Dashboards, a web app that can help organizations collate and monitor trends.

As Hockett updated her knowledge of GIS technology, she turned to a colleague, Sunshine McDonald, who works for

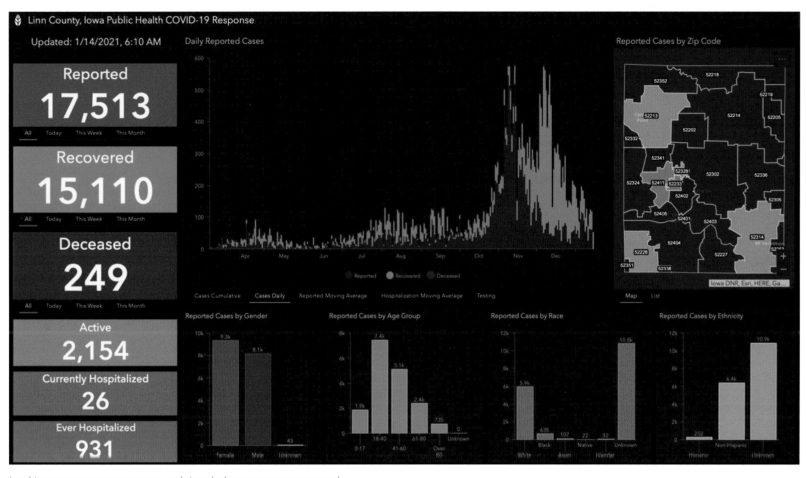

Looking at cases across space and time helps everyone see trends.

the nearby Marion Police Department and has experience processing enormous amounts of data and then plotting the results on maps.

With advice and help from McDonald, Hockett was able to layer many categories of data so residents in need of food aid could simply enter their addresses on the map to find nearby public agencies and private charities. Many of Linn County's newest residents come from cultures with restrictive diets for religious or cultural reasons, so mapping their location and needs helped avoid food waste and ensure appropriate supplies.

The use of dashboards and maps prompted others to suggest adding new layers of data—and made Hockett and other officials aware of still more unmet needs. This was especially crucial when the windstorm hit.

"Due to the derecho, we didn't have electricity for almost a month," Hockett said. "And we had a lot of people whose electricity was crucial for at-home medical equipment."

Once the vulnerable people were identified and mapped, first responders checked on them. That required help from power companies, medical groups, and the public at large. And it involved connecting a lot of data, which was no easy task. But it was a necessary task, and that's all the motivation Hockett needed.

"We built relationships early on, and people know to come to us for support now. Going through a derecho and COVID-19 has really helped illustrate the power of GIS and what we can do," Hockett said.

Chile Achieves Rapid Vaccinations with Dashboards Marking Progress

Chile has achieved one of the quickest distributions of COVID-19 vaccines in the world. By mid-April 2021, more than 50 percent of the population had received at least one shot, well beyond any other Latin American country and behind only a few countries in the world.

Chilean epidemiologists credit this rapid response to both the country's network of clinics—placed even in the remotest regions—and its modern information infrastructure that receives input from every community.

"We incorporated the concept of territorial intelligence to share data with partners and the general population so they can make better-informed decisions," said Cristián Araneda, executive secretary of the Chilean National System for the Coordination of Territorial Information (SNIT).

During the pandemic, SNIT gained stature because it spans all ministries as a permanent mechanism for institutional coordination. A regular meeting led by Julio Isamit, minister of National Assets, convened to discuss the status of COVID-19

across all territories and to share Ministry of Health data. A team of analysts created the country's COVID-19 case index and vaccination index to analyze and present this data to the various ministries to gauge impacts and guide policy.

SNIT also worked to share COVID-19 updates with the public. The team first created the COVID-19 dashboard to provide a way to see and share infection rates. They quickly pivoted to track progress on vaccine distribution with the Yo Me Vacuno ("I get vaccinated") dashboard.

"The first thing we asked ourselves was why, what is the purpose of the dashboards?" Araneda said. "Our goal is keeping the public informed on progress and giving citizens a tool to make decisions. We want to be transparent and improve the quality of life. We also want to share good news, such as with our elderly population, who are so happy after getting vaccinated, they feel as if they have another chance at life. Now in Chile, the population has the same information that the president of Chile and the minister of health have to make decisions about the pandemic."

Establishing Territorial Intelligence

SNIT, also known as IDE Chile, is the spatial data infrastructure organization that established a data exchange mechanism and optimized the management of geospatial information in Chile. This includes many datasets on people, energy, the environment, transportation, natural resources, government property, and disaster response. Every dataset that SNIT manages contains the foundational element of location for analysis and visualization as layers on a map using a GIS. Now that SNIT has aggregated all the data, they spend more time on analysis to generate knowledge.

"Our territorial intelligence policy is intended to get us closer to the population," Araneda said. "We are very technical, but we know that maps can tell a better story than raw data and present the information in a way that makes it easy for the general public to absorb and understand."

With the COVID-19 crisis, SNIT helped the Ministry of

National Assets become a trusted source for public guidance, allowing everyone to see infection rates as well as vaccine management and distribution. The COVID-19 dashboard alone has received more than 4 million visits in this country of 17.5 million people. The dashboard approach has been successful with the public, politicians, and the media.

"Today, it's not uncommon to see the president making decisions using the COVID-19 dashboard or talking about the dashboard on television," Araneda said. "Every week we can meet with famous doctors on TV presenting information and drawing conclusions with the dashboard in the background."

For every effort, SNIT applies great rigor to represent the data correctly and make the tools easy to understand.

"We run a validation step with people who are not GIS specialists to make sure users can understand the details at a single glance," Araneda said. "I also want to emphasize the importance of our multidisciplinary approach to gain a variety of perspectives and to avoid reworking a project at the final stage. We have a well-tuned development process, like any R&D center, that allows us to achieve quick implementation."

Vaccination Campaign

In preparation for vaccinations, SNIT analysts first used GIS to determine the location of vaccination sites. The vaccination team could look at sites and data on a map to consider such things as how to reach the most people, and to make it easy for the public, including factoring the location of metro stops.

Next, the team created the Yo Me Vacuno dashboard to inform the public about the progress of the vaccination campaign while providing a map to guide people to vaccination hubs. It took SNIT just 10 days from the idea of the vaccination dashboard to receiving and processing the data to launching it for public use.

"Showing the vaccination hubs helps the public make fast decisions," Araneda said. "It gives people official up-to-date information without confusion, so people can go to their nearest vaccination center, for example."

Public sign shows symptoms of COVID-19.

They also built a dashboard with voter registration information and demographics for Chile's Constitutional Convention election in May 2021, when voters elected governors, mayors, city councilors, and members of the assembly to draft a new constitution. This dashboard is popular, and the public has come to expect informative and interactive maps.

"Without a doubt, awareness of GIS has grown during this crisis," Araneda said. "Various government departments and nongovernmental organizations are getting in touch with us to set up more systems because they now understand how important these systems, maps, and dashboards have become."

Agencies and other groups are discussing ideas for future dashboards, such as systems for firefighters, environmental issues, and maps for children. SNIT's efforts have been recognized both inside and outside the country.

"Recently, the ambassador of Peru had a meeting with the minister, and one of the main points of the conversation was how to replicate the dashboard in Peru based on the Chilean experience," Araneda said.

Resurgence of Cases Provides Lessons for Everyone

The country's vaccination efficiency remains a marvel, but like most first movers, Chile has learned lessons others should heed. A recent rise in COVID-19 cases indicates that Chileans

Measuring response to the Yo Me Vacuno dashboard has been done in a more qualitative than quantitative way. Rather than views of the dashboard, SNIT has been tracking public sentiment.

"We have seen comments on social media and our website, such as 'Thanks to this dashboard I have been able to get a vaccine quickly and safely,'" Araneda said. "People are taking the time to let us know it has had a positive impact on their lives."

Sharing National Information

During the pandemic, the SNIT team put together many dashboards on other topics such as which national parks or small businesses are open.

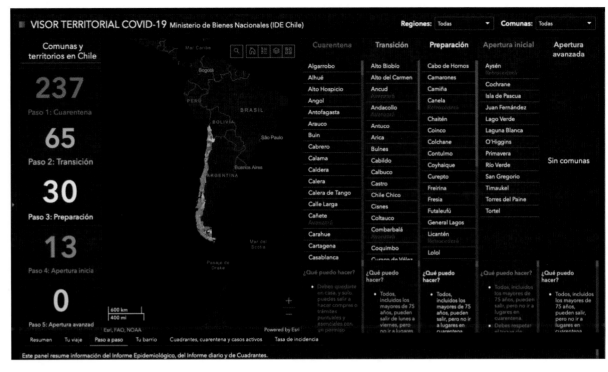

The COVID-19 dashboard in Chile is regularly shared on television because it provides a way to see and share infection rates.

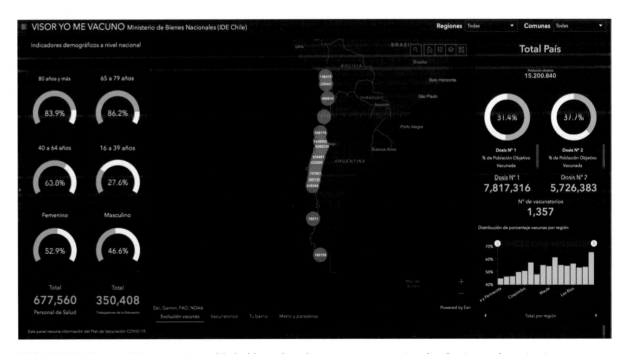

Chile's Yo Me Vacuno ("I get vaccinated") dashboard tracks progress on vaccine distribution and vaccination rates.

may have let their guard down too soon.

"The situation is very fluid," Araneda said. "A month ago, we were very happy and claiming success, and now we are in quarantine again."

Chile had been restrictive, with closed borders between March and November 2020. When cases dropped to less than 1,000, and to help a faltering economy, Chile opened its borders to international travelers for the holidays and summer season in the Southern Hemisphere.

Officials are attributing the latest surge to many factors, including new COVID-19 variants and the easing of personal protection steps such as masks and frequent handwashing.

"Right now, it's very difficult for us to have a strategic long-term plan," Araneda said. "Like before, we continue to have consistent meetings to look at data and make improvements to our response."

The COVID-19 Delta Variant: This Map Provides Answers

By Este Geraghty, Esri Chief Medical Officer

The United States experienced its fourth wave of COVID-19 infections over the second half of 2021 resulting from the far more infectious Delta variant of the virus. The mutation with a coinciding spike in cases was expected, but the rampant nature of the spread across the country posed many questions. First and foremost is, where do I need to be especially careful?

The interactive Esri map *Which Way Are Things Going?* gives a county-by-county view of active cases for a week-by-week perspective on COVID-19 trends.

Southeastern states saw surges, led by Louisiana, Mississippi, Arkansas, Georgia, and Florida. Whereas rates were lower in Iowa and South Dakota, these two neighboring states were places to watch after the Sturgis Motorcycle Rally in August 2021 caused concern as a potential super-spreader event. The map is dynamic, allowing viewers to track the impact of significant events as they unfold.

The map aims to provide accurate and actionable information for everyone, with data from state level to county level that makes it easy for anyone to zoom in and see the trends that will help them make decisions about when and where to better safeguard their health and that of their family. Despite the Centers for Disease Control and Prevention (CDC) guidelines renewing the call for indoor masks where cases

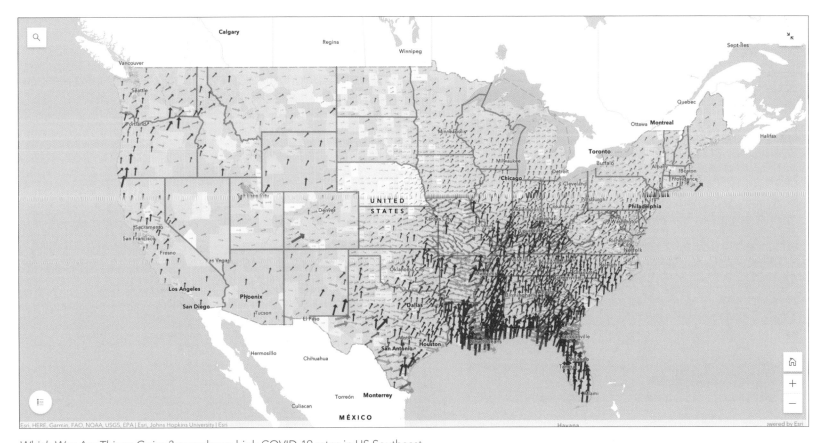

Which Way Are Things Going? map shows high COVID-19 rates in US Southeast.

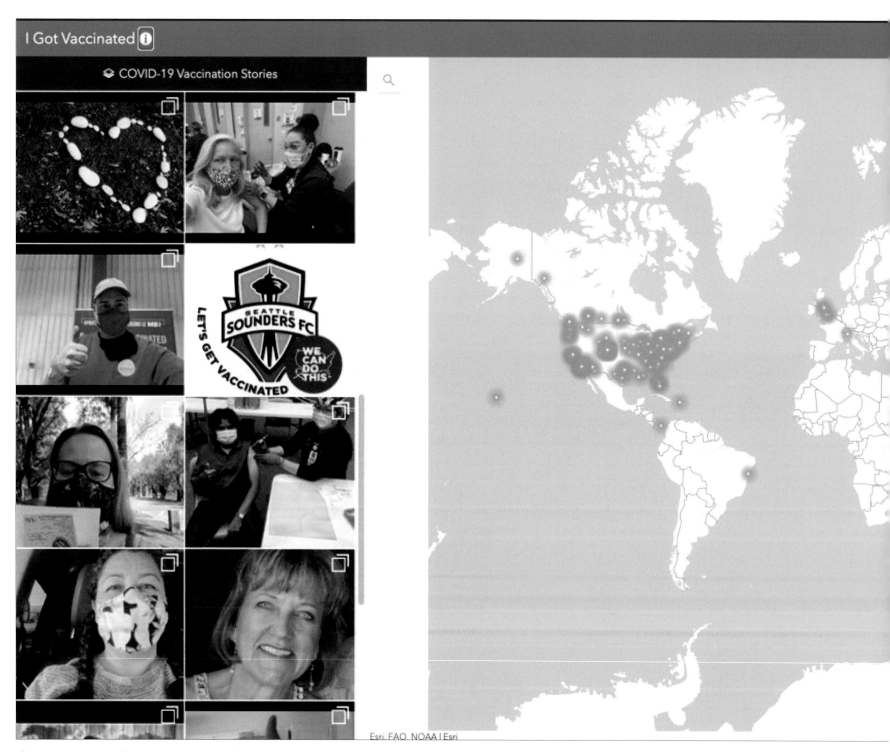

The *I Got Vaccinated* StoryMaps story provides a means for people to share their vaccination stories.

were highest, many people dialed back on personal and business plans even as reopening continued.

The Ongoing Need for Geographic Pandemic Information

During the many waves and phases of the COVID-19 pandemic, a map-based view has proven compelling and informative—letting people know whether things are getting better or worse in their area. Johns Hopkins University (JHU) set an early example for COVID-19-related map-based insights. Data from that effort also underlies this weekly map update.

The original Johns Hopkins COVID-19 Dashboard has become an iconic source of information. It has surpassed all prior metrics for an online issue-oriented map with trillions of views to date. And the open data approach the team took has facilitated the work of other countries, local governments, and businesses from around the world as they present a localized view of the data to their constituents and stakeholders.

A serendipitous alignment of technical knowledge, a desire to inform, and the work ethic to harvest data from around the world helped the JHU team become a trusted and authoritative source of information in the crisis. The background on how this map originated has continued to fascinate people even as eyes turned from case counts to how and where to receive a vaccine.

What's Next for COVID-19 Mapping?

At a high level, we may feel like we're all in this together. And we are. There is a level of interconnectedness across the globe that inspires solidarity as we all continue pandemic response efforts. At the same time, we know that the COVID-19 pandemic's impact is not the same for everyone.

Communities experience the virus and its consequences unevenly and often unfairly. People of color, essential workers, and those with poor access to health care and services are at higher risk of contracting the virus. These and other determinants related to COVID-19 can be mapped and analyzed in a way that supports resource allocation equitably and fairly.

We also know that the per capita rate of new cases follows a pattern—the highest case rates occur where the fewest people are vaccinated, compounded by geographic variations in adherence to other public health measures such as masking and social distancing. Mapping can support ongoing messages of vaccine efficacy and protection from the worst symptoms and most devastating outcomes of the virus.

We can encourage broader vaccine uptake by sharing individual stories, such as the crowdsourced US I Got Vaccinated map. Anyone can contribute their vaccination experience. Mapping techniques are also used to better understand unvaccinated populations in ways that help health officials more precisely support their varying needs from place to place.

Scientific and spatial analysis will continue to outline the patterns and anomalies arising from the ongoing global spread of this outbreak, and mapping will aid epidemiologic understanding of those patterns and targeted actions.

As we all anxiously wait for this crisis to pass, maps will continue to point the way to better public health preparedness and response.

It's in the Wastewater: How UC San Diego Senses and Maps COVID-19

Before the Delta variant became commonplace, toppling any optimism that the COVID-19 pandemic might soon be over, a team at the University of California (UC) San Diego got an early glimpse at the new threat. There it was, the virus's RNA sequence, in their wastewater samples.

Rather than cause alarm, the presence of the more infectious COVID-19 variant offered the team hope. Here was evidence that their epidemiological strategy could detect the Delta variant and likely any future mutations.

To communicate cases, effluent testing results are tied to a live map created with a GIS to relate which buildings have had a positive reading. The virus sensing effort has driven early testing and helped curb campus spread. It's also enabled UC San Diego to continue to offer on-campus housing and in-person classes and research opportunities throughout most of the pandemic.

Although other campuses have attempted similar screening programs, few have reached UC San Diego's scale or efficiency.

(Courtesy of UC San Diego)

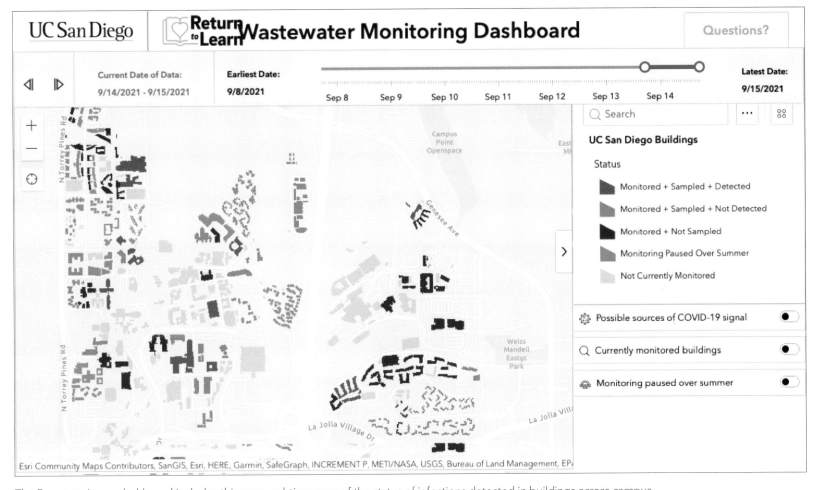

The Return to Learn dashboard includes this near real-time map of the status of infections detected in buildings across campus.

To be successful, the system must provide near real-time information. Now, the same team has been tapped to monitor wastewater for the entire city of San Diego's 2.3 million residents using the same innovative process.

How UC San Diego Led its Return to Learn

The campus setting has proven to be an ideal place to devise disease surveillance systems and curb the spread of COVID-19 despite people living and working in close quarters.

UC San Diego's Return to Learn program was initiated with three pillars in mind: risk mitigation, viral detection, and intervention. The university's Operational Strategic Initiatives department spearheaded the effort, with Rob Knight, professor and director of the Center for Microbiome Innovation, as the chief laboratory investigator.

Concerned that outbreaks could occur and quickly spread in highly occupied buildings, Knight and his team initiated wastewater testing as a way to safely maintain campus population during the pandemic and take early action when needed.

Often, infected students weren't aware they had the virus. In the summer of 2020, the system detected a positive case one Friday afternoon. Notifications went out within 14 hours, and that weekend more than 650 people were tested.

"We had several examples where we could see that it was a single infected individual in a building of several hundred students," said Natasha Martin, an associate professor in the Department of Medicine, Division of Infectious Diseases and Global Public Health at UC San Diego. "When we found that person and moved them to isolation, the wastewater was negative the next day."

The Return to Learn team credits transparency as the key to participation. Detailed maps of every building, pipe, and sewage access point form the basemap. A public GIS dashboard shows daily updates of buildings that have been monitored, sampled, and detected. This keeps the entire campus community abreast of the university's viral infection status.

"It's not just our processing of the data but being able to share that data out to everyone and the public," Martin said. "Even now, we have people that visit, and they can check on the public dashboard and see if a particular building was positive on that day."

A student gathers wastewater from the autosampler. The next stop is the lab.

Despite early uncertainty about how well the system would detect cases, it proved to be "incredibly sensitive," Martin said. More than 85 percent of the university's residential cases were detected in the wastewater.

Showing Students, Staff Where the Virus Is

At the outset, UC San Diego had just six wastewater samplers that captured effluent for laboratory testing. The results were recorded on a spreadsheet manually. Now, UC San Diego samples the wastewater in 350 buildings every day, and the results automatically get added to the map.

"The hours of me sitting in front of a map and an Excel spreadsheet, trying to crosswalk all of those signals with the buildings, are thankfully over," Martin said.

To achieve automation, UC San Diego had to make a better map that included the pathways of the sewer systems so they could correlate wastewater samples collected in pipes to the buildings they served. Once a particular building was identified, they could trace the virus to individual people through the standard nasal-swab testing.

During the summer of 2020, the Return to Learn team began issuing notifications to on-campus residents and staff by email, lobby fliers, and even door-to-door knocking to compel further testing. To the surprise of Martin and others, it worked. When people received notifications of positive wastewater samples, individual testing rates jumped as much as 13 fold. The testing prompted a quick start of isolation and contact tracing.

Automation Involves Robots, Dashboards, Maps, and Listservs

To alert people on campus of a positive result, the team couldn't rely on a time-consuming multistep lab process. Backlogs and delays would have derailed the promise of wastewater testing. To speed up the process, UC San Diego introduced robots to the lab and automated the notification systems.

A group of students and staff gather every morning to collect sewage samples using liquid-handling collection robots across 350 campus buildings. Then, they return to the lab for processing. In the lab, robots concentrate the virus using magnetic nanoparticles, and then extract RNA—the genetic material that makes up the genomes of viruses such as SARS-CoV-2—from the samples. Polymerase chain reaction testing is then used to search for the virus's signature genes.

The automated, high-throughput system can process 24 samples every 40 minutes. Data is then added to a digital dashboard and map that tracks new positive cases building by building.

If Martin knows a student is already isolating in the building in which a positive sample comes back, she doesn't issue a notification.

The notification process is otherwise fully automated. Martin provides the dates of positivity and the sampler number, and that automatically issues emails to everyone residing or working in the building.

She estimated that automation initially saved her two to three hours per day. Without automation, it would have been hard to expand the scale of the wastewater testing effort.

Expanding Beyond the Campus

With the promise of curtailing future epidemics and pandemics, wastewater surveillance could be a university's or community's greatest defense without the need for more intrusive methods of surveillance. The US Centers for Disease Control and Prevention has recognized its efficacy, which has been validated and refined during the pandemic, and a new National Wastewater Surveillance System was launched.

"As the barrier to entry and [operating] continues to drop, we hope wastewater-based epidemiology will become more widely adopted," Knight said in a press release. "Rapid, large-scale infectious disease early alert systems could be particularly useful for community surveillance in vulnerable populations and communities with less access to diagnostic testing and fewer opportunities to distance and isolate— during this pandemic, and the next."

How Maps Inform Public Health Decisions in San Bernardino County

The Research, Assessment, and Planning (RAP) team at the San Bernardino County Department of Public Health (SBCDPH) in California monitors the health and well-being of more than two million residents in the country's largest county by area. Already a big job before the COVID-19 pandemic, the work became more challenging when office

shutdowns and stay-at-home orders affected all 25 programs that deliver services to people facing heightened needs. In addition to continued support of the programs that rely on daily analysis, the RAP team also developed one of the state's first COVID-19 tracking hubs using GIS technology.

The county also understands the value of a geospatial lens for public health programs beyond tracking the pandemic. One recent project investigated the reason for a nearly decade-long decline in participation in the county's Women, Infants, and Children (WIC) program. Analysts created a smart map showing how and how far people must travel to reach their local WIC clinics, enabling officials to determine better placements for the sites to improve accessibility.

The initiative is part of a larger countywide goal to be more transparent both internally and with the public about how data informs decisions. Lap Le, statistical analyst at SBCDPH, explained how his team uses geospatial technology to analyze people and places and communicate the decision-making process.

"We have all this data, and it's the primary driver in terms of an evidence-based approach to making decisions to positively affect the community," Le said. "But at the same time, we also need to provide the data so that the public can understand what is driving those kinds of decisions."

Solution

A Dispatch Tool, consisting of a *Web Map* with food facilities, inspector assignment, and status layers *(assigned, not assigned, visited, not visited, etc.)* as well as the Batch Attribute Editor widget was created with *ArcGIS Web AppBuilder*. Using this editor, supervisors could select clusters of facilities and quickly assign them to an inspector via the facilities' attribute table.

The Research, Assessment, and Planning team compiled and communicated relevant metrics for vital decision-making regarding the health and safety of San Bernardino County residents to the Department of Public Health executive team and generated a dashboard to communicate case numbers to the public.

An Evidence-Based Approach

For more than 35 years, the WIC program in San Bernardino County has provided families with access to healthy foods, nutrition education, and community resources as part of the larger federal program. Since 2013, the number of program participants among eligible residents has been steadily declining, with a sharp drop in participation starting in January 2020 because of the COVID-19 pandemic.

When Le and his team began their WIC analysis, they started by interviewing clinic staff who believed one of the key barriers to participation was people's ability to easily get to the clinic either by foot or public transportation.

Le explained, "At the ground level, when you talk to the clinic staff, they already understand where the needs lie. But we needed some solid evidence to justify certain moves or relocate clinics."

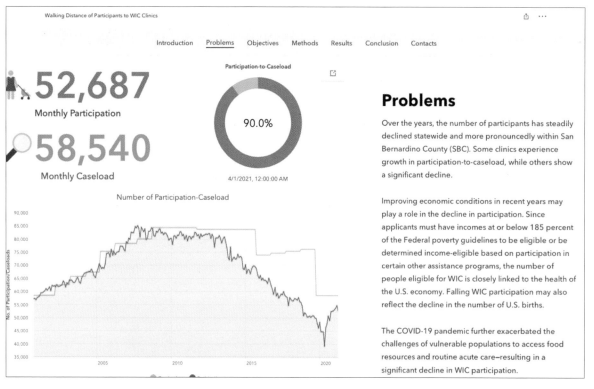

A story highlights the problem of reduced participation in the WIC program, alongside strategies and actions.

Team members decided to map where current and past program participants lived throughout the county in relation to the clinics. They used GIS to display the aggregated data on a map. This allowed them to quickly understand where most of their participants lived without revealing street addresses or other confidential data.

Serene Ong, GIS analyst II for SBCDPH, explained that the next step was designed to test the hypothesis they had gathered from clinic staff. "Then we thought, 'Okay. Is it easy or difficult for these participants to get to our clinics?' So we created visual buffers on the map that indicated half-mile and one-mile walking distances around each clinic." They also added a map layer that displays all the bus routes in the area to understand whether participants who lived farther from each location had the option to take public transportation to their appointments.

Ultimately, the county used the information to tailor decisions about where WIC clinics should be located to best serve the needs of the community. "Our analysis is what drove the decision to move some of the clinics to a denser location, and then remove certain clinics where there's less of a need," Le said.

Ong hopes that the next phase of the project will include mapping the density of potential participants for the program so that WIC can do a better job of reaching communities that need these services.

COVID-19 Accelerates Public Health Digital Transformation

Both Ong and Le joined SBCDPH in 2019, shortly before the COVID-19 pandemic began. One of their first projects together was creating an interactive dashboard to share information about the spread of the virus in the county.

As with many other local governments throughout the US, county officials were initially getting daily updates via a PDF document. The analysts used GIS to aggregate the information, pulling data from the California Department of Public Health's Reportable Disease Information Exchange (CalREDIE) database to automatically update the dashboard daily. The team supplements state metrics with county-specific data.

"When the dashboard was first published, other counties wanted to have discussions about our workflow, how we update the data, and how we structure all the components," Ong said.

Since the beginning of 2020, the dashboard has grown and changed significantly to better serve its users. "We get a lot of feedback both internally as well as from the public, so we're always improving and making innovations to the dashboard to make it more user-friendly and interactive. It's our goal to keep making it better for everyone," Le said. One example is the addition of a school tab where users can view what's going on in each school district.

Through their experience creating and optimizing the COVID-19 dashboard, team members have been able to apply lessons and best practices to other data projects within the county. They've created dashboards for various programs and taught teams how to interact with the data themselves. "Now, instead of having each of the program stakeholders making

requests, we can anticipate needs and have the data readily available," Le explained.

The county uses smart maps and dashboards for a range of projects—from addressing vaccine hesitancy to improving efficiency for inspectors in the field. According to Le, county officials were initially hesitant to display any data visually. "Eventually, we made a few maps to show them how useful it is in terms of data display and representation instead of just using tables and charts. When it's convenient or when it's doable, we always use a map to represent our data, just to make it easier to interpret. Initially, there was a lot of pushback."

Le paused and then added, "But we convinced them otherwise."

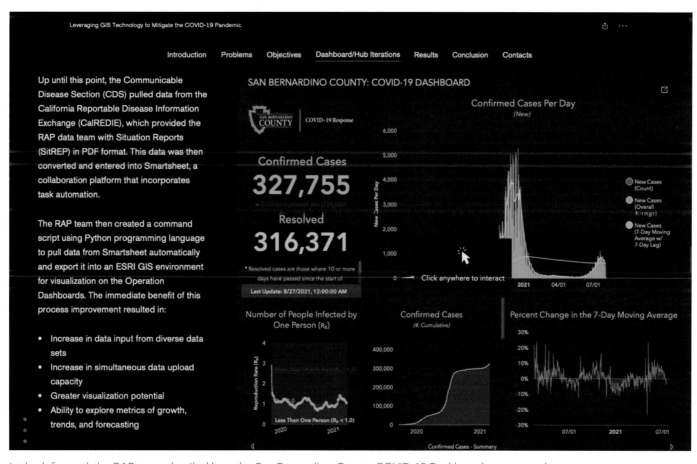

In the *left panel*, the RAP team detailed how the San Bernardino County COVID-19 Dashboard was created.

Equity and Social Justice

Race and place are good predictors of whether people can thrive, so geography is central to addressing inequities. US president Joseph Biden recognized this when, on the day of his Inauguration, he signed the Executive Order on Advancing Racial Equity and Support for Underserved Communities. This order pinpoints the need to assess the geographic location of underrepresented or disadvantaged communities and manage programs using technology.

In the social sphere, policies that address equity and social justice issues use GIS to measure and map diversity and direct investments to communities that have been neglected. Local governments have been pioneers in this approach, and state and federal efforts are following.

GIS provides the means to assess inequities and act.

Smart maps that display US Census Bureau demographic data about age and income, layered with information about senior living facilities, homeless encampments, hospitals, and clinics, have become vital to health-care professionals, government officials, and community leaders during the COVID-19 pandemic. Map views give immediate insight into where the number of serious cases may spike and where supplies or health-care infrastructure should be bolstered.

The Social Vulnerability Index from the Centers of Disease Control and Prevention (CDC) uses a composite of 15 social factors to flag communities that might experience a punctuated demand for health care that outpaces the local medical infrastructure. When exploring age and other social vulnerability factors across US locations that correlate with

higher susceptibility to COVID-19, health and city officials and local emergency responders can see where an influx of health needs will likely occur.

Today, experts have repeatedly established the link between environmental and equity issues. Communities of color disproportionately bear the consequences of excess pollutants and contaminated soil and water, including higher-than-average rates of disease ranging from asthma to cardiovascular diseases to cancer, as well as birth defects and other health disorders. Scientists, policy makers, and activists often lay bare the connection by using GIS technology, creating interactive fact-based maps to bring the issue into stark relief.

The rise of the environmental justice movement has coincided with the increase in the capabilities and accessibility of GIS software that can process, aggregate, and display these connections on smart maps. As GIS has become more commonplace, it has also become integral to thousands of environmental justice projects, strengthening arguments for racial and social equity.

During the pandemic, epidemiologists used GIS to understand the spread of COVID-19, while environmental justice advocates used the technology to communicate related response disparities.

This innovation may ultimately prove to be how GIS can most help the environmental justice movement, by underscoring the unnatural aspects of the impact of natural disasters and development decisions.

If More Women Owned Land, More People Might Be Fed

By Jen Van Deusen, Esri Sustainable Development Industry Solutions lead

Since the first seeds of civilization were planted, access to arable land has been central to human life. Yet even today, 12,000 years later, ownership of rich farm soil has largely been denied to one gender. Women, who make up half the global adult population and 43 percent of the agricultural workforce, account for less than 15 percent of farmland owners. Nearly 40 percent of the world's economies still limit women's property rights, according to the World Bank. In the remaining 60 percent, malecentric cultural norms and legal inefficiencies still hinder women's rights.

Ensuring women have the right to own land could lift millions of people out of hunger, reduce rural poverty, and improve sustainable natural resource management, according to studies from the United Nations (UN), World Bank, women's rights nonprofit Landesa, and others.

Such sources also agree that achieving global gender equity will start with local efforts—often by putting more information in the hands of women.

"It's about data empowerment," said Amy Coughenour Betancourt, CEO of Cadasta, a global nonprofit that connects communities with technologies for recording land and resource rights. "We approach it as enabling local stakeholders to document, secure, and manage their land rights and helping communities understand and use the data that's collected."

Cadasta's tools are built on ArcGIS technology, which allows users to digitize local land records; view property parcels on a map; and more easily find information such as land use, legal ownership, or value. With transparent accessible data, women can better understand and exercise their rights to land.

Information Agency: Data Gives Women the Power to Choose

The systems that govern individual rights to own, hold, or use land are frequently a combination of national laws and regional customs.

Under some systems, for example, a married woman's legally inherited land—or a married couple's jointly owned land—is considered the customary sole property of her husband. Widows, divorcees, or unmarried daughters could similarly be expected to cede control of their property to a male relative.

About one in five adults worldwide live with the worry that their land could be taken away from them, according to a 2020 study by Prindex, a research group that tracks global perceptions of land rights. Although the study found that men and women were nearly equally likely to feel insecure in their landownership, women were up to 21 percent more likely to cite divorce or spousal death as a possible cause of losing their land.

"Women are often workers, but they don't get to be involved in decisions about which crops to plant, where to plant them, or what kind of agricultural practices to use," Coughenour Betancourt said. Women are then left with little agency to support themselves or their family absent a male head of household—contributing to cycles of poverty and food insecurity.

Giving women the choice to take ownership of their land

Cadasta team works with partners at PRADAN and local forest-dwelling communities in Odisha, India, to map individuals' land boundaries. (Photo credit: Cadasta Foundation and PRADAN, India)

starts with building a better understanding of their legal rights, and that begins with capturing digital records. Better access to maps and property data can also simplify proof of ownership, which can be expensive and time-consuming to obtain, particularly in rural communities. In Mozambique, for example, Cadasta worked with local leaders and the National Cooperative Business Association Cooperative League of the United States of America (NCBA CLUSA) to make it easier for local farmers to get a certificate of landownership. Digital maps created with GIS technology eliminated several surveying steps, lowering time and costs from upward of a year and US$400 in some cases to as little as three months and $30.

"Mozambique is a really good example of how the technology facilitates these processes," Coughenour Betancourt said. "And the efforts in Mozambique emphasize women-led farms—women are given priority, and if the woman is married, she's included on the certificate along with her husband."

Coughenour Betancourt notes the importance of creating formal records and getting women's names documented in landownership. When they aren't included in writing, women are particularly vulnerable to losing their homes or livelihood.

Where Equity and Efficiency Meet

Within the last decade, the UN has identified land and housing security, the role of women in agriculture, and women's land rights as key indicators in meeting Sustainable Development Goals (SDGs) covering poverty, hunger, and gender equality. Most of the world's poor and hungry are women and girls. Focusing on women's rights targets an especially vulnerable population.

Evidence also shows that the benefits of gender equity in land and resource distribution extend far beyond women's empowerment, uplifting not only women but entire communities. Women lose an estimated 20 to 30 percent of farming capacity because of limitations placed on their access to land, tools, and education. With equal access, women could produce enough food to lift between 100 million and 150 million people out of hunger.

Local community group meets with a paralegal team while working with Cadasta's partners at the Council of Minorities in Bangladesh. (Photo credit: Council of Minorities, Bangladesh)

Mozambican women and their families receive a government-issued land right certificate based on data collected on Cadasta's platform
under NCBA CLUSA's conservation agriculture program, PROMAC. (Photo credit: NCBA CLUSA Mozambique)

In addition, Coughenour Betancourt said, "Women inherently invest more in their families with their income," referencing a collection of studies from around the world that show improved outcomes in health, food access, and education when women are included in family landownership. Women farmers are also more likely to grow crops that can feed their families and support local commerce.

At a time when COVID-19 has disrupted supply chains and thrust millions of people worldwide into food insecurity, leaders are looking for ways to bolster sustainable food systems that keep communities fed when global distribution fails. Evidence shows that women's access to land and education can help fill the gaps by boosting local food production and growing nutritious, resource-efficient crops.

"A sustainable food system has many essential components," Coughenour Betancourt said, "three of which are land rights, climate-smart agricultural practices, and women's empowerment. And one of the best ways to promote women's empowerment is to give a woman a right to land."

Land to farm has long provided for basic needs such as food, shelter, and income. Beyond survival, though, land rights can be a gateway to a person's financial independence, social mobility, and political inclusion. This cycle of opportunity is central to the sustainable future that the UN SDGs enable. Transparent, accessible, and equitable systems empower more individuals and, in turn, create stronger communities that are better equipped to withstand the challenges of our time.

Online Schooling Prompts Municipalities to Map Digital Inequities

When the coronavirus pandemic upended in-person classrooms, children were adversely affected. Adults fortunate enough to retain jobs they can do from home are at least familiar with the concept of remote work. For kids, live video classwork reflects an alien pedagogy, and the lack of in-person peer interaction has hindered the development of social skills.

Pivoting to computer-based remote learning has also deepened existing fault lines in American education. With broadband integral to schoolwork, researchers at the Pew Research Center have documented a widening homework gap.

Teens in households earning less than $30,000 are nearly three times as likely to report having trouble completing homework assignments, because of lack of access to a computer or reliable broadband connection, compared with households with incomes above $75,000. Nearly half of teens in low-income families say they sometimes do their homework on a cell phone.

Pew also found that the homework gap involves racial disparities. More than one in five Black teens are forced to search out public Wi-Fi sources for connectivity. Hispanic teens

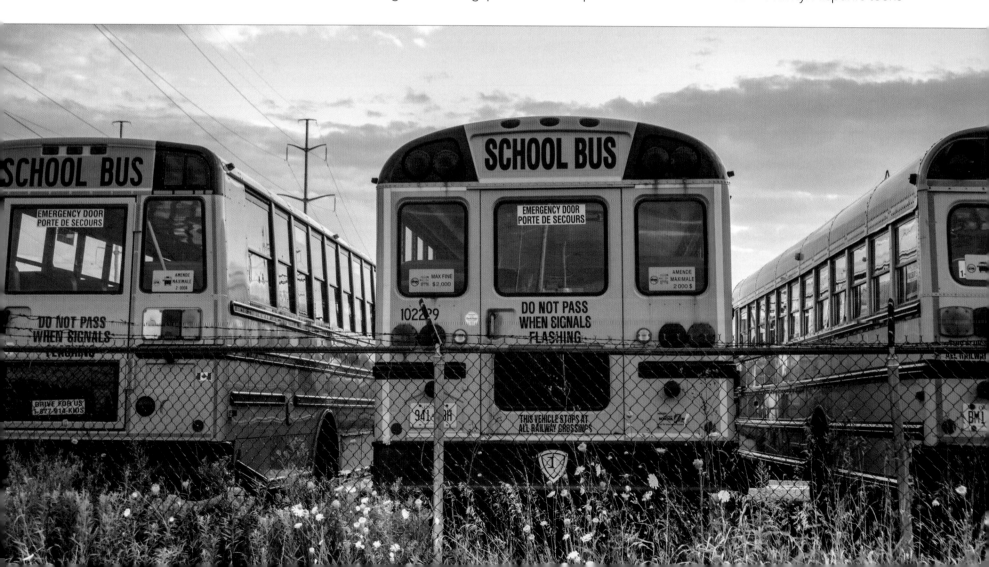

Students had to pivot to video conferencing software for in-home learning.

are twice as likely as White teens to report they lack access to a home computer.

The pandemic made the problem impossible to ignore. "The digital divide was suddenly starkly apparent," said Mark Wheeler, chief information officer for the City of Philadelphia's Office of Innovation and Technology (OIT). "There were families who couldn't afford reliable internet. They were using whatever they could to get by—public computing centers, free Wi-Fi through businesses, libraries—and with those shuttered, we had a sizable population that couldn't participate in daily life."

The school closures that roiled American life in spring 2020 underscored a crisis within a crisis. Almost overnight, school districts, especially those with large populations of students from low-income families, had to devise ways to keep the gap from swallowing students whole. Here, we look at what has happened in two of the largest school districts in the US.

From Philly to Florida, a Common Goal

At first glance, the city and county of Philadelphia would seem to have little in common with southeastern Florida's Palm Beach County. Each has a population of around 1.5 million, but Philadelphia's is squeezed into an area one 16th the size of its distant neighbor on the tropical end of the Eastern Seaboard.

Philadelphia has twice the poverty, two-thirds the median income, and significantly less sunshine.

Scratch the surface, however, and Philadelphia and Palm Beach look more alike. Although Philadelphia is one of America's poorest large cities, it contains pockets of high affluence. Although Palm Beach County's considerable wealth is mostly concentrated on the Atlantic coast, in cities such as Palm Beach and Boca Raton, poverty persists in several communities sandwiched between the ocean and the Interstate 95. The poverty rate in the Glades, the inland area near Lake Okeechobee, is higher than Philadelphia's, and the median income is much lower.

Philadelphia and Palm Beach also share a civic quirk. Both counties consolidate their schools into a single school district that each serves around 200,000 students. One of every 130 public school pupils in the country lives in one of these two districts.

The arrival of COVID-19 forced the school districts of Philadelphia and Palm Beach County to quickly determine who was most at risk before the school year ended.

Palm Beach Rises to the Challenge

Each district acquired 80,000 Chromebook computers, using a combination of philanthropic and public funds. To understand better who most needed these computers and how to ensure these students and others also had broadband access, both districts adopted similar data-driven approaches.

To assess needs, the Palm Beach County schools' Division of Performance Accountability sought the advice of Donna Goldstein, an IT manager with the district. Goldstein's area of expertise is GIS, software that analyzes people and places based on location-specific data.

Goldstein began her analysis by using geocoded student data to see where students lived. The data contained demographic information relating to each student, which could be displayed as layers on a map showing where incomes were lowest. She could also see who had already accessed the online student portal.

Goldstein's team was especially interested in students coded as eligible for subsidized school lunch, thinking these might be families with limited access to technology. Using GIS, she performed a "point density analysis" that displayed geographic clusters of families, which Goldstein then color-coded into three levels of concentration.

These clusters helped the county devise where to place Wi-Fi hot spots, pole-mounted transmitters that broadcast Wi-Fi signals that students nearby can access with a special receiver.

To refine the data, Goldstein added more layers, including municipalities, census tract data, and the location of neighborhoods already earmarked for community revitalization funds. She also mapped housing subdivisions because the county would need to request easements to install the receivers on private property.

The map also helped the county plan where to lay fiber to bring the signals to the Glades. To reach far-flung homes, officials used the map to pick buildings on which to mount poles, including schools, a church, a library, and an Animal Care and Control office.

"We combined all this data to give us a really well-rounded view of what's going on in the county, and where the greatest areas of need were," Goldstein said. "If the team wasn't able to work with the data geographically, and had to look through raw databases, I don't know how long it would've taken, but it would've been an extraordinarily long time."

The wealthy people of Palm Beach, Florida, mostly live on this narrow isthmus along the Atlantic coast.

Bridging the Digital Divide in the City of Brotherly Love

Meanwhile, a similar process was unfolding in Philadelphia. A program called PHLConnectED, a joint effort launched by the OIT and the Mayor's Office of Education, was helping families obtain computers and establish home Wi-Fi hot spots, while also making plans to establish community computing access centers. By late October 2020, more than 11,000 public school families were receiving free internet access, thanks to the collaborative effort of city government, the school district, and business and philanthropic leaders.

As in Palm Beach, the initial difficulty that PHLConnectED faced was how to identify families that needed the program the most. Those experiencing housing insecurity were hard to contact. An even bigger problem was finding families in relatively stable situations but with limited incomes. "We have to make them realize what we offer them is free, and that this isn't a rug that will be pulled out from under them after a few weeks," Wheeler said.

To help organize efforts, Wheeler's office tapped CityGeo, a dedicated team within OIT devoted to mapping and spatial analysis. CityGeo was already using GIS to maintain a city "stress index" that compiles geographic data on crime, homelessness, drug abuse, and other issues that would suggest the existence of students in need. The data helped PHLConnectED prioritize the distribution of wireless routers to create mobile hot spots for students.

"A lot of our work is focused not only on mapping, but on keeping data dynamically up to date through the dashboard," said Hank Garie, CityGeo's geographic information officer. "So whether it's meal sites or access centers, it's all fed into the GIS, which gives us a great way to visualize and analyze where we might want to target outreach programs based on need or affordability."

Post-COVID, Closing the Larger Gap

The progress made around bridging the homework gap in Philadelphia and Palm Beach County has implications for broader equity issues. "As spinoffs from this initiative, we've been able to do parallel work with our Commerce Department," Garie said. "A lot of these same datasets are applicable, and we can even use them to view the city's budget through an equity lens."

Wheeler noted frequent references to the "stress index" in city meetings as the spatial analysis "really brings into stark relief that so much of this is about where people live."

"The work the CityGeo team has done has really laid the groundwork for me, as a CIO, to have conversations at the mayor's level about where populations we're trying to reach live, and how they're aligned with so many other critical problems we're trying to solve in the pandemic," Wheeler said.

Goldstein agreed. "For me, it goes beyond kids," she said. "That's our primary focus, but now you've got parents and other adults in the home who have broadband access, which opens up whole new worlds of possibilities for them economically."

She sees the effort as providing a small silver lining during the crisis. "From my perspective, this is one of the only good things to come out of the pandemic," she added. "As educators, we've been fighting the digital divide for eons. So this is really exciting."

With schools closed in Philadelphia, a program called PHLConnectED worked to provide computers and internet connections to disadvantaged students.

Maps Show the Real Picture of Race and Equity in Oakland

Recently ranked as one of the most racially diverse cities in America, Oakland, California, celebrates and protects its diversity. Yet, like many other major US cities, Oakland has a long history of systemic racial and economic discrimination. In the 1940s, federal and local government policies destroyed traditionally diverse neighborhoods and segregated the population.

True to their reputation, Oaklanders fought back. Oakland was one of six cities that led a general strike in December 1946, resulting in the modern labor movement. Community

groups born in the 1960s—including the Black Panther Party, Oakland Community Organizations (PICO/OCO), The Unity Council, and Intertribal Friendship House—demanded protection and equal access to jobs, housing, employment, transportation, and services.

Oakland continues the fight for racial equity today. In 2015, it became the first city in California to start a Department of Race & Equity, focused on creating a community in which diversity is not just maintained but celebrated, racial disparities are

identified and eliminated, and equity is achieved. To reach these goals, the Department of Race & Equity uses innovative tools and technologies, including a GIS to help every City of Oakland department ensure that the work it's doing advances racial equity.

Generating a Baseline for Understanding Inequalities

The Department of Race & Equity is a small team of three people led by its director, Darlene Flynn. Before her move to Oakland in 2016, Flynn helped create the Race and Social Justice Initiative in the City of Seattle's Office for Civil Rights. That's where she developed a theory of change that is the backbone of the department's approach to promoting equity in Oakland. Flynn's theory starts by acknowledging that race matters, disparities in most indicators of well-being are race driven, and to achieve social justice, barriers must be removed for all racial groups.

The approach focuses on promoting equity as an internal organizing structure with the city proactively working to dismantle long-standing and current racial disparities. All departments of the city government have a role to play by examining the outcomes of their policies, practices, and procedures for underserved communities and identifying actions to address the disparities and advance racial equity through individual and collective actions geared to create systemic change. Oakland's Department of Race & Equity provides the messaging and analysis approach as well as technical tools and support.

One of the tools the department is using to do this work is location intelligence, which comes from a GIS platform. According to Jacque Larrainzar, who worked on Flynn's team in Seattle and is now an equity program analyst in Oakland, GIS is essential to the work they do since geography has played a huge role in the way racial discrimination and disparities affect Oaklanders. "Structural racism is actually written in our geographies, especially in zoning and planning codes, but also in transportation," she said.

That's why, when the Department of Race & Equity was assembled in 2016, one of the first things staff did was secure a grant from the City University of New York (CUNY) Institute for State and Local Governance for research to understand how racial disparities in Oakland impact its residents. With the support of CUNY, the department analyzed 72 different indicators of well-being and disaggregated the data by race for resources including housing, education, and health care. They used GIS to organize and analyze existing datasets to examine the disparities between the most and least impacted groups by racial disparities across geographies and within six themes associated with the standard of living: economy, education, public health, housing, public safety, and neighborhood and civic life. The output of that work was the 2018 Oakland Equity Indicators Report.

In addition to a published report, the team displayed results on smart maps where stakeholders could easily visualize how and where inequities existed throughout the city in ways that are not immediately apparent with charts and graphs. Larrainzar explained the impact this had on members of the community. "It not only helps people see the disparity but also helps them to think about how to make it better,

Graffiti declares civic pride for the city.

which I found is really hard for folks because these things are embedded in everything we do. Which makes visualizing disparities something different, hard sometimes. Those visual juxtapositions really help people to see what is invisible, discuss how we can go about changing it, and how implementing certain changes actually promotes equity."

Fighting Disparities, One Project at a Time

The results of the Equity Indicators Report were mixed. Oakland scored relatively high in matters of civic engagement, but scores for public safety were dismally low. Overall, the city came out with a score of 33.5 out of 100. Department staff weren't surprised by the results and viewed them as a data baseline to understand how things can improve over time. With that baseline in place, the department has embarked on some ambitious, equity-focused projects with its city collaborators.

Let's Bike Oakland!

One of the department's first projects was a revision of the city's bike plan, in partnership with the Oakland Department of Transportation (OakDOT). For this endeavor, the team pulled data from public records and used geospatial analysis to identify Oakland's most vulnerable groups in terms of access

The diversity of Oakland is on display during a Black Lives Matter protest.

to protected bike lanes, public safety, commute times, and road conditions. Then, OakDOT partnered with community organizations led by people of color, based in east and west Oakland, to ensure that those historically left out of these conversations and impacted by racial disparities were prioritized and heard in the bike planning process via community listening sessions and design workshops. GIS helped the team collect and organize data and prioritize where and when bike lanes would be added based on need.

Equity-Focused Paving Plan

OakDOT is in the middle of a three-year equity-focused paving plan guided by the Department of Race & Equity. Larrainzar explained how the team used GIS to determine which streets would get paved during this project and the importance of using an equity-focused approach for projects such as this. "When we start talking about implementing equity, GIS is very helpful because we can actually go down to the street level. We looked at how many people traveled on a street. Who lived around that street? What facilities were around that street that provided services for these communities? And we started to get a very clear picture of what streets needed to be prioritized and maintained because we didn't have enough money to pave the whole city."

Smart maps generated by GIS also helped gain the support of other residents. The OakDOT team would show residents the condition of their road and the number of residents it served versus some of the higher-priority areas with roads in far worse condition that served 10 to 20 times as many people. Residents throughout the city were able to see the evidence for decision-making in a compelling way.

2030 Equitable Climate Action Plan

Oakland recently passed the city's new 10-year plan for mitigating and adapting to the climate crisis. The goal was to prioritize the needs of residents most vulnerable to the effects of climate change. In addition to an 18-month community engagement process, including workshops, social media, climate equity workdays, an online survey, and town halls, location intelligence was considered heavily in decision-making. Displaying data geographically allowed stakeholders to easily see a correlation between the number of trees in a neighborhood and the rate of asthma among residents, for instance.

"You need to be data-driven, and you need to really be based on reality," Larrainzar said. "A lot of times, governments are very siloed. They tend to work just with the people that work in the city, sometimes even in just one department."

For the Oakland team, GIS plays a key role in gathering resident feedback so the city's plans will improve lives.

"We might be experts on this particular field of work, but if we don't go back to the community and really hear from them whether this is going to be helpful, we might not help them," Larrainzar said. "We might create something that's worse for them."

Mapping Equity during a Pandemic

In March, when COVID-19 started to pose a serious problem, the Department of Race & Equity decided to test a theory that the virus may be impacting people of color more because of the impacts of structural racism. Staff worked with officials in Alameda County to become one of the first counties in the country using GIS to track the impact of COVID-19 on different racial groups.

When they looked at the data on race and geography on maps, it became clear. Eight zip codes were inordinately affected by COVID-19, with the Hispanic/Latinx community the most impacted by far. This information prompted additional social services for those residents as well as the launch of Oakland's Flex Streets Initiative, which makes it easier for retail stores, restaurants, and other businesses to use larger portions of the sidewalk, parking lanes, and streets for open-air interactions and proper social distancing.

To choose the right locations for Flex Streets and offer the right services for affected residents, the team engaged community feedback through a texting app. All suggestions were input into a GIS dashboard where they were displayed on a map and analyzed to find the solutions that best meet the needs of residents.

"I was surprised that I needed street-level understanding of where to put facilities and services and resources based on who was there," says Larrainzar of what she's learned during the pandemic. "When I work on data reports, we typically look at blocks, and we leave it there. But really, what I'm finding is that when we're talking about designing services, you really need to get to that level. And mapping is very helpful for that because if you have the right data linked to those mapping apps, then you can very easily start designing real solutions for the folks who live there."

After four years, the Department of Race & Equity is seeing clear results. "I tell people the analogy of moving a mountain, like, we are actually moving the mountain," says Larrainzar.

Oakland's mayor, council directors, and city staff are taking notice. In each department, people are viewing their efforts in terms of equity—whether it's planning and zoning, building, public works, trash pickups, street cleaning, park maintenance, or programs for youth.

The technology supports and even inspires this level of interest, according to Larrainzar. She said: "GIS allows us to connect with folks that have no technical background and to have real policy-making discussions from a very different place than if you just went in there and showed them a PowerPoint with all the data and the pie charts."

Oakland leaders are engaging with the community to ensure that people who have been historically left out are heard.

How Austin's Map of Trees Helped City Leaders See and Tackle Social Inequities

In Austin, Texas, like many places, the number of trees in neighborhoods marks a divide of race and income.

For Austin, the correlation can be seen in tree canopy maps that city staff have overlaid with demographics and other data using a GIS. In west Austin—the area west of Interstate 35—tree canopy covers 78 percent of the land. In east Austin, tree canopy covers only 22 percent.

"It's really interesting that Interstate 35 is also a dividing line for ecoregions," said Alan Halter, a senior GIS analyst with the City of Austin. "If you go west, you get into the Hill Country, with a lot more tree cover, but east Austin hasn't historically supported as many trees. And when you look at who lives where throughout Austin's history, communities of color have resided in east Austin."

Inequity in Numbers of Trees

In 1928, Austin's Master Plan (a term no longer used because of its racial connotations) relegated the city's Black communities to a district east of present-day Interstate 35. Redlining made it nearly impossible for residents to move, whereas it placed fewer restrictions on White residents to purchase homes in Austin's heavier-canopied parts of town.

In the 1950s, environmental injustices were locked in when

the planning commission zoned all east Austin property as "industrial," affecting nearby residents with the area's lower air quality, higher temperatures because of a lack of tree cover, and other health-related issues.

This common pattern is found throughout the world—the prevailing winds, blowing west to east, bring pollution to the eastern parts of town.

When Halter first mapped Austin's trees, he was focused on the city council's urban forest plan.

ECOREGIONS

EDWARDS PLATEAU   BLACKLAND PRAIRIE

I-35

The ecoregion divide shows a sharp disparity in tree cover, noted in green.

"I created the first Community Tree Priority map back in 2015, and it was really tree planting oriented—to figure out where to plant trees," Halter said. "At that time, equity was a consideration but wasn't really a main focus. We were mostly looking at where tree canopy existed and didn't exist, with the idea to increase shade across town and get trees where they're not currently located."

As Halter added layers of data to the map, he saw the relationship between socially vulnerable neighborhoods and areas with minimal trees. In the wake of Black Lives Matter protests, equity became a strong focus of the Austin Community Climate Plan, so the map needed to change.

"We released an update in 2020 with equity as the driving force," Halter said. "We're now looking at tree planting to achieve positive outcomes for people, such as improved public health; reduced heat island effects; and, of course, addressing climate change, because it's related to everything."

To understand the impacts of having fewer trees, the city participated in an urban heat island mapping project coordinated by the federal government's National Integrated Heat Health Information System, a collaboration between the National Oceanic and Atmospheric Administration (NOAA) and the Centers for Disease Control and Prevention (CDC).

"Volunteers drove different routes in their cars with fancy devices poking out of their windows that recorded temperatures every few seconds," Halter said. "Using GIS, we could extrapolate temperature readings on a larger scale to see what heat looks like around town and compare it to tree canopy."

The result was an interactive web map showing that morning temperatures were higher in dense urban areas close to the city center than in other areas. Results exposed how concrete structures that absorb solar heat in the day and radiate it at night can be seven degrees hotter than outlying areas in the

East Austin (*left*) has few trees whereas West Austin's suburbs (*right*) are full of trees.

day and five degrees hotter at night. Strategies to mitigate the effects of urban heat islands include white roofs, more crosswalks so that people don't have to walk as far, more bus shelters, and, of course, more trees.

The Effect of Trees on the Lives of Residents

Planting more trees in an underserved area starts a positive chain reaction: more trees means more canopy; more canopy means more shade; more shade means less heat; less heat means lower energy bills and more outdoor activity. Therefore, more trees results in improved health and quality of life.

Trees create fresh air while also cleaning some pollutants. The greenery is appealing, which draws people outside where they can move around and be more social. A recent study even found that street trees are present where fewer people take medications to deal with depression. Trees also actively cool areas in a process that's similar to perspiration.

"The scientific term is *evapotranspiration*," Halter says. "I noticed it on a superhot day in July at midday, and suddenly these trees started to cry or sweat, as if it was raining. The trees are taking up water from the ground, then it goes up to the leaves, and then the tree rains on itself and the water goes back into the soil. It's kind of a breathing, liquid-to-gas process."

This analogy is fitting, since forests are often called the lungs of the earth, but most people don't experience the process so directly.

"Trees cool the environment—you can actually feel it," Halter said.

This measurable benefit is often referred to as an ecosystem service.

Trees also help protect areas from increasingly severe storms—especially important in a place such as Austin, which experiences frequent cycles of drought and flooding. Tree roots draw in rainwater and keep the soil from washing away. The leafy limbs slow heavy raindrops before they hit the ground, so the soil is less prone to erosion.

Selecting the Right Trees

In 2014, the US Forest Service (USFS) conducted an inventory of trees in Austin to help understand tree canopy in detail and assess the carbon sequestration capacity of trees.

USFS analysts determined that in Austin, there are currently 33.8 million trees, which store about 1.9 million tons of carbon dioxide. Researchers found that every year, the trees remove about 92,000 tons of carbon as well as 1,253 tons of air pollutants and reduce residential energy costs by $18.9 million.

The inventory included a species review, finding that the most common trees are Ashe juniper, cedar elm, live oak, sugarberry, and Texas persimmon.

Ashe juniper, Halter said, is "the number one cause of tree allergens in Austin—but it's also the tree species with the greatest air quality impact because it's an evergreen species, and Ashe juniper trees act as year-round air filters. Ashe juniper also captures the most storm water runoff and sequesters the most carbon in our urban forest. It's a weird dichotomy—the tree that's disliked the most is also the tree that's helping us the most."

Like all urban forestry specialists, Halter and his colleagues in Austin have a difficult management task dealing with weather and infestations. Oak wilt, a fungal disease that is spread by beetles and gets into the root system, can be spread from tree to tree underground. The emerald ash borer is a nonnative pest that hasn't hit yet, but Austin is getting ready because once it's present, it typically wipes out the entire ash tree population.

"We don't yet know the impact climate change will have on pests like borers or the fungus causing oak wilt, but we can expect things to get worse with a rise in temperature, which increases the stress on trees," Halter said. "We have tree doctors that go out and give shots to the trees for things like oak wilt, but it's really tough for trees to respond and survive if they're not getting enough water to begin with."

Trees, Equity, and Social Justice

According to Austin's climate plan, the city "recognizes historical and structural disparities and a need for alleviation of these wrongs by critically transforming its institutions and creating a culture of equity." The city gives away free trees every year, but it has only recently examined—from an equity perspective—where those trees have been planted.

"We have to look at who has the means to drive across town on a Saturday to pick up trees," Halter said. So trees are now offered at giveaway events located in previously underserved neighborhoods. "With canopy mapping, we can assess it and show people, through the data, what that looks like. And the Community Tree Priority map—with scoring metrics to show areas of higher need—focuses our grant funding and a lot of tree planting."

To help with this work and greater community outreach, the city has established the Youth Forest Council.

"We offer paid internships, which provide our youth a pathway to green careers," Halter said. The students in the council work alongside professionals in Austin's urban forestry program, gaining practical knowledge about the natural and environmental sciences and using GIS for urban forestry. Ultimately, the Youth Forest Council helped shape the Community Tree Priority map, providing valuable input about trees, equity, and health. Halter hopes the students start to see how GIS helps people understand complex relationships.

"Of course, we know that climate solutions have the potential to improve the quality of life for all people," Halter says. "But we also know that climate change impacts don't really affect everyone equally.

"And that is at the heart of our plan. We are preserving existing trees and planting new trees, where trees are most needed."

The targeted efforts of the Community Tree Priority Map aim to address past inequities.

Maps and Data Strengthen Calls for Housing Equity in Houston

As the floodwaters of Hurricane Harvey subsided from Houston, Texas, in 2017, they revealed a troubling pattern: low-income, Black-majority neighborhoods bore the brunt of the destruction.

Houston's long history of devastating floods devalued the areas in and around floodplains—and as a result of bygone segregated housing practices, communities of color have been established largely on this underresourced, flood-prone land. Houston remains demographically split along these same geographic lines, with persistent disadvantages affecting the quality of housing—and quality of life—for many Black residents.

A recent report offers a visualization of disparities between Houston's project-based Section 8 apartment complexes in Black-majority versus White-majority neighborhoods. Based on a two-year study published in January 2020 by the nonprofit housing advocacy group Texas Housers, the report maps and compares housing conditions and opportunity indicators. The findings align with Houston's historically divided landscape—and the maps, built with GIS technology, help clarify the correlation between race, geography, and social inequity.

"Urban policy is often trying to explain concepts to people. Maps provide a clear visual—here is rich, here is poor; here is Black, here is White," said Mia Loseff, a community equity analyst at Texas Housers.

Loseff and the team at Texas Housers hope the report will give residents an evidence-based resource as they continue to fight for safe, equitable housing.

Quantifying Quality of Life

Stories of Houston residents living with Harvey's aftermath provide a glimpse into the inequities ingrained in the city's terrain. Even three years later, devastating details surfaced after Texas Housers staff interviewed residents of Houston's predominantly Black Section 8 apartment complexes. Jamie Wasicek, a resident at Coppertree Apartments, says the mold in her unit has triggered breathing difficulties for her five-year-old son. He occasionally misses school as a result. Daija Jackson, a resident at Arbor Court Apartments, lost everything during Hurricane Harvey—the complex is built directly in a floodway. Jackson was displaced during repairs and returned to find the flood damage barely rectified.

Meanwhile, the team's photographs from site visits to Section 8 properties in the Woodlands, a northern suburb, paint a different picture. Residents of Copperwood, Holly Creek, Wood Glen, and Fawn Ridge Apartments enjoy amenities such as manicured landscaping, units with private balconies, and community pools and playgrounds.

There's no question that these close-up accounts are powerful examples of the report's findings. But the numbers tell a more sweeping, equally compelling story: in Houston, more than 90 percent of low-income renters are Black. In the Woodlands, that number falls to less than 20 percent.

Additionally, about 25 percent of the Houston complexes are in high-risk floodplains. The Woodlands properties are all in areas with a minimal history of flooding.

Finally, all but one of the complexes located in Houston score in the bottom quarter of nationwide crime indexes—their neighborhoods are rated more dangerous than 75 percent of the country. The Woodlands apartments score in the safest 50 percent.

"The subsidized housing program is supposed to promise opportunity and high quality of life for everyone," Loseff said. "This particular program does have the potential to work well, but in Houston, it hasn't for people of color."

On a map, it's easy to see why there's such a stark contrast in residents' experiences. Houston's race and wealth maps are nearly identical, with a clear, arrow-shaped patch of White, high-income residents on the east side of downtown. Forty of Houston's 43 Section 8 apartment complexes are located outside

Houston, Texas, is often referred to as the Bayou City because it's crossed by a number of swampy rivers that impact lower-income communities when they flood.

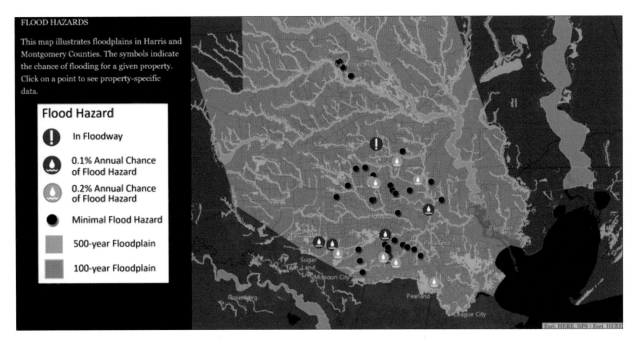

Flood Hazard

⚠ In Floodway

💧 0.1% Annual Chance of Flood Hazard

💧 0.2% Annual Chance of Flood Hazard

● Minimal Flood Hazard

◼ 500-year Floodplain

◼ 100-year Floodplain

Texas Housers map shows areas in Houston with frequent flooding. The dark dots show locations of Section 8 apartment complexes.

parents and grandparents still frequently limit the ability to afford a college education, secure a high-paying job, or invest in their children's future.

Housing—and the location of that housing—is a critical turning point in this cycle. "People often look to surrounding census tracts as a predictor of how successful you'll be," said Loseff. "The impacts of raising a family in some of these neighborhoods can alter the positive growth trajectory of a child's life."

this area in low-income, Black- and Latinx-majority communities.

Along with income, flood risk, environmental hazards, and crime rate, the report also maps Real Estate Assessment Center physical inspection scores and high school scores. In all categories, the Houston apartments fare worse than their counterparts in the Woodlands.

"These maps tell a clear story," said Loseff. "They reveal two separate and unequal living experiences, divided across lines of race. You see the history of a nation that has just refused to integrate low-income people into mixed-income locations."

Opportunity Costs

Lately, as the US contends with entrenched social injustices, the term *systemic* has become a key component of the equity conversation. Part of the fabric of American society, systemic inequities are a remnant of outdated mind-sets and overturned policies that have yet to fully fade into the past. Today, for example, although Americans of color are legally free to pursue the same opportunities as their White contemporaries, socioeconomic adversities faced by their

In 2013, for example, economist and Harvard professor Raj Chetty mapped social mobility in the United States, finding that for low-income families, geography had a clear causal effect on a child's chance at financial betterment. The Texas Housers report corroborates Chetty's work: even from one low-income family to the next within a 40 mi. span from Houston to the Woodlands, location determines access to essential developmental resources such as good schools, outdoor activities, and environmental and personal safety.

But, the report also suggests, systemic factors often determine locations where people can live, making higher-opportunity areas such as the Woodlands all but inaccessible to certain groups. As with many other cities in the US, Houston's communities of color still reside in the less safe, less resourced spaces they were given under eras of segregation, and moving out isn't always on the table.

The Houston metro area has just 19 affordable and available housing units per 100 low-income people, according to the 2018 The Gap: A Shortage of Affordable Homes Report.

And this statistic doesn't account for the quality of those units. As evidenced by the locations of Houston's Section 8 complexes, even when residents can choose between apartments, it's often the choice between one underresourced neighborhood or another. What's more, with project-based housing, the Section 8 subsidy is tied to properties and landlords, not tenants.

"If a tenant chooses to move [out of these complexes entirely], they will lose their subsidy—the very tool that makes housing affordable to them," said Loseff.

Inner-city residents also establish support systems among family, friends, and neighbors. "A lot of people are really hesitant to leave an apartment," said Loseff. "It's 'My gram is down the street, my neighbor can watch my kids.' That network will tie them there."

Although a safer, affordable home could lie just a half hour away in the Woodlands, changing location can mean starting over with jobs, childcare, and cultural acceptance—costly in both real and social currency.

"We base much of our advocacy on four basic rights for homes and neighborhoods," said Loseff. The first of these rights is the ability to choose freely where to live in a decent, affordable home. As the report points out, *choose* is the operative word: until safe housing in high-opportunity areas is a viable option for all low-income renters, there is still work that needs to be done.

The Incredible Importance of Home

From *Homer's Odyssey* to *The Wizard of Oz*, the concept of home has long been linked to comfort and stability. But approximately 8 percent of Americans receiving Section 8 assistance may lack this sense of safety, based on data from the 2017 American Housing Survey. Respondents categorized their living conditions as severely or moderately inadequate, based on a variety of physical criteria such as functional plumbing, visible mold, and pest control.

In Houston specifically, Loseff said, "It creates cumulative disadvantage if you have to live in some of these properties that perform so poorly. We have to recognize this and decide to do better."

With the release of this report, Texas Housers also called for reforms to protect Houston's low-income renters. Recommendations include raising standards for Section 8 funding, better enforcement of the Fair Housing Act, building new complexes in higher-opportunity neighborhoods, and including residents in decisions about maintenance and upkeep.

Although the first three items rely on policy makers and regulatory bodies, the resources that Texas Housers supplies to low-income Texans can help give them a stronger voice in the housing conversation right away. Among those resources are GIS maps depicting the current situation.

"For people who live in these underfunded neighborhoods, maps help confirm what they already know and experience," Loseff said. "They understand that they're living in an area that's not as nice as the wealthiest place in town, but with these maps and reports, it's backed by data." The goal is to help residents make a stronger case to landlords, council members, or housing authorities, and to empower them to know and fight for their right to a safe home with equal opportunity.

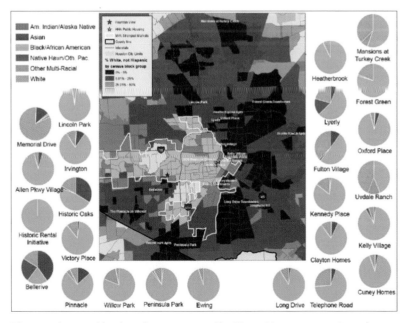

This race by neighborhood map, created by Texas Housers, examines the racial profile of Houston's neighborhoods.